THE COLOUR CODE

THE COLOUR CODE

PAUL SIMPSON

PROFILE BOOKS

First published in Great Britain in 2021 by

Profile Books
29 Cloth Fair, Barbican, London EC1A 7JQ.

www.profilebooks.com

1 3 5 7 9 10 8 6 4 2

Typeset in Adobe Garamond Pro
to a design by Henry Iles.

A CIP catalogue record for this book is available from the
British Library.

ISBN 978-1781256268
e-ISBN 978-1782832423

Printed in Italy by L.E.G.O. S.p.A. on FSC® paper

CONTENTS

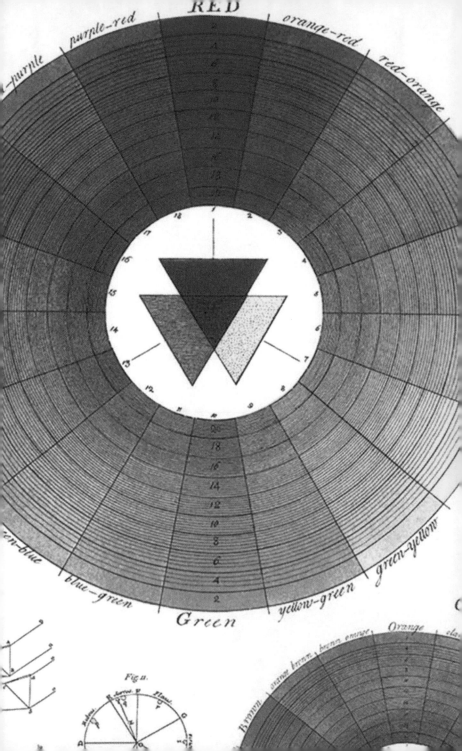

INTRODUCTION

'Colour directly influences the soul. Colour is the keyboard, the eyes are the hammers, the soul is the piano with many strings.'
Wassily Kandinsky

How many colours are there in a rainbow? The obvious answer, ever since Sir Isaac Newton codified the spectrum, is seven: red, orange, yellow, green, blue, indigo and violet, which gives us the acronym ROYGBIV. But in his treatise *Meteorologica*, Aristotle suggested that there were just three principal colours in the rainbow: red, green and purple. The appearance of yellow, Aristotle argued,

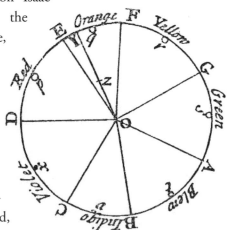

was merely an effect of the contrast of the red against the green. Anthropologists suggest that for the Pirahã and the Candoshi peoples of the Amazon, who have no specific colour terms in their language, the rainbow has only two tones: darker/cooler and lighter/warmer. In reality, there is no particular number of colours in a rainbow, because each colour blends imperceptibly into the next. When we give names to colours, we are imposing an order on the small part of the electromagnetic spectrum we call 'visible light' (wavelengths of c.400–740 nanometres).

In deciding that seven was the correct number, Newton – who admitted that 'my own eyes are not very critical in distinguishing colours' – might have been drawn to the age-old pattern of sevens (seven days in a week, seven wonders of the world, seven notes in the musical scale, seven liberal arts, etc.), and to the mystical aura with which the number had been invested by Pythagorean philosophers.

In his book *Opticks* (1704), Newton categorised colours as primary (red, blue and yellow), secondary (green, orange and purple) and tertiary (hyphenated colour names). If you mix the primary colours, you can create every other colour. His experiments proved that white light could be separated into pure prismatic colours ,which could then be combined to make white light again. As he concluded: 'If the Sun's Light consisted of but one sort of Rays, there would be but one colour in the whole World.'

Newton's analysis was not universally popular. John Keats famously lamented that Newton had 'Destroyed the poetry of the rainbow by reducing it to a prism', while German polymath Johann Wolfgang von Goethe made a passionate case for colour as a subjective, rather than purely scientific, phenomenon in

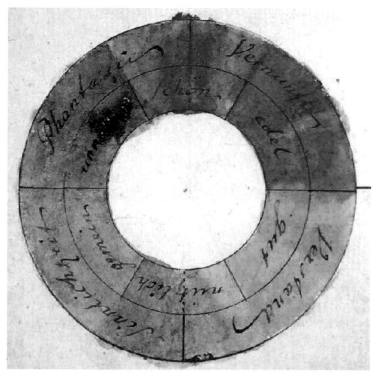

Goethe's symmetric colour wheel with 'reciprocally evoked colours' (1810).

his book *Theory of Colours* (1810). Colour, Goethe argued, was created by the interaction between the physical behaviour of light and the apparatus with which we perceive it. Accordingly, he divided the spectrum into life-enhancing 'plus' colours (yellow, yellow-red) and anxiety-inducing 'minus' colours (blues, purples and blue-greens). As philosopher Ludwig Wittgenstein observed: 'What Goethe was really seeking was not a physiological but a psychological theory of colours.'

Goethe's insistence on the emotional force of colour inspired J.M.W. Turner, who acknowledged his debt explicitly in the title of his painting *Light and Colour (Goethe's Theory) – The Morning after the Deluge – Moses Writing the Book of Genesis* (1843). Over time, Goethe's ideas would be embraced by artists as diverse as Vincent van Gogh, Konstantin Malevich, Wassily Kandinsky (whose book *Concerning the Spiritual in Art* reflects Goethe's influence) and Mark Rothko.

It could be said that in emphasising the subjective element of visual perception, Goethe was a forerunner of thinkers such as the French cultural historian Michel Pastoureau, who has written a series of brilliant books on colour. We see the world through a prism more complicated than Newton's – a prism in which our emotions, culture, age, gender, religion, politics, sporting allegiance and personal experiences all come into play. As Pastoureau puts it: 'Colour is, first and foremost, a social construct.'

Goethe's reflections on contrasting, complementary and successive colours were formulated more scientifically by French chemist Michel-Eugène Chevreul. In 1824, Chevreul was tasked with reviving the fortunes of the Gobelins tapestry workshop in Paris. Customers had complained that its colours were too dull and grey. After investigating the dyes, which were as bright as anyone else's, he concluded the problem was not chemical but optical – the apparent dullness was caused by the way that the colours were juxtaposed. Chevreul's consequent law of simultaneous contrast was elaborated in his book *The Laws of Contrast and Colour* (1839), in which he systematically analysed the ways in which the intensity of any colour was affected by an adjacent one. Bringing all the colours in the visible spectrum together in a wheel, he showed that complementary colours – opposites on the colour wheel – packed more visual punch when juxtaposed.

His book became the most widely used and artistically influential colour manual of the nineteenth century. Eugène Delacroix was so convinced by Chevreul's work, he declared he could 'paint the face of Venus in mud, provided you let me surround it as I will'. The Impressionists recognised that by applying brushstrokes of pure colour to a canvas – and letting the viewer's eye combine them optically – they could make light and colour more brilliant. One of

A plate from Chevreul's *The Laws of Contrast of Colour* (1839).

Chevreul's colour effects, the use of massed monochromatic dots, inspired the Pointillism of Georges Seurat and Paul Signac. The abstract colours employed by the Orphists – particularly Robert Delaunay, Sonia Delaunay and František Kupka – are also rooted in the chemist's seminal work.

How do we see colour? Through the optic nerve, the brain receives signals from two kinds of light sensors at the back of the retina – rods and cones. In essence, rods enable us to see us in dim light, while cones enable us to distinguish colour in bright light. The nineteenth-century English scientist Thomas Young proposed

A plate from Thomas Young's *Lectures*, published in 1807, showing his grasp of ocular anatomy and the wave theory of light.

that our cone cells are sensitive to three wavelengths: red, green and blue-violet. The German physicist Hermann von Helmholtz developed this theory, arguing that each cone perceives light in one of these wavelengths and the relative strengths of these wavelengths are interpreted by the brain as colour. Most of us are trichromats because we have three types of cones, each of which can see 100 shades. So, the possible number of colour combinations our brains can see is one million. A much-disputed number of (usually female) tetrachromats have four cones, and can thus see one hundred million hues.

One in twelve Caucasian men are deuteranopes – 'red–green' colour-blind – compared to one in twenty Asian men, one in twenty-five African men and one in 200 women. The inability to distinguish between blue-yellow and blue-black is much rarer. Colour blindness is a genetic trait carried on the X chromosome which is usually compensated in women by the second X chromosome.

A 2006 study by biologists at Cambridge University and the University of Newcastle tested the idea that people who cannot tell red from green have a different kind of light receptor in the eye that is more sensitive to other hues. They asked people to rate the similarity of fifteen circles painted in khaki tones. Those with regular vision struggled. Deuteranopes distinguished the tones easily, leading researchers to conclude that they could see a different dimension of colour.

Most mammals are dichromats, so they can only see 10,000 colours. Some – including humans, some primates and, recent research suggests, many marsupials – are trichromats. One theory, proposed by Robert Finlay in his 2007 paper 'Weaving the Rainbow: Visions of Colour in History', is that trichromatic mammals, wary of becoming a dinosaur's lunch, became nocturnal and traded a cone for a rod, becoming dichromats, because being able to see more clearly in dim light was more useful than distinguishing between colours. After the extinction

The multisensual, multicoloured mantis shrimp has up to sixteen sensors that identify colour. Butterflies have at least five. Most humans have three. Dogs just two.

of the dinosaurs, some mammals developed a third cone to help them identify food and, it has been suggested, interpret situations by recognising, for example, that red skin may signify anger. Many birds are tetrachromats, having an additional photoreceptor which can see UV colours. Butterflies have at least five receptors. The peacock mantis shrimp, found in the Pacific and Indian Oceans, has up to sixteen types of sensor in its eyes.

American science journalist and broadcaster Robert Krulwich once created a small furore by declaring that pink was an artificial colour, on the grounds that no single wavelength of light looks pink. It's true that pink is a mixture of red and purple light, but to suggest that pink is therefore not a 'real' colour is to fundamentally misunderstand what colour is. As biologist Timothy H. Goldsmith argued in *Scientific American* in 2006: 'Colour is not actually a property of light or of objects that reflect light. It is a sensation that arises within the brain.' The light-sensitive cells (photoreceptors) in the eye detect wavelengths of light within specific ranges and at particular locations. This information is dispatched through the optic nerve to neurons in the early visual cortex, which interpret the information to create a picture. In the past, we assumed that colour and shape were processed separately in the early visual cortex and combined later, but a 2019 study by the Salk Institute in California, using the latest imaging technology, suggests that they are encoded together. Around 40 per cent of the brain is said to be involved in processing visual information, but neuroscientists still do not understand in detail how it performs this task.

'Colour is the place where our brain and the universe meet.'
Paul Klee

The complex neuroscience of colour is vividly illustrated in Oliver Sacks' essay 'The Case of the Colorblind Painter'. An artist identified as Mr I lost his ability to see colours after a car accident at the age of 65. He told Sacks: 'My vision was such that everything appeared to me as a black and white television screen. My vision became that of an eagle – I can see a worm wriggling a block away. The sharpness of focus is incredible. But I am totally colour-blind.'

Trapped in a world where everyone looked like 'animated grey statues', Mr I lost his appetite because every dish looked black. His psychological recovery began when he changed his external world to match his perception of it, eating black olives and white rice, drinking black coffee, and becoming nocturnal because the world looked more natural to him at night.

One morning, out driving, Mr I saw the sun rise. To his eyes, the blazing reds of dawn were all black 'like a bomb, like some enormous nuclear explosion'. Realising that no one had ever seen a sunrise in quite that way, he painted it, in black and white. Mr I became so proud of his vision – and his art – that when a specialist told him he could retrain his brain to see colours again, he found the idea repugnant.

After much study, Sacks concluded that two parts of the brain are critical to our understanding of colour. The cells in an area of the visual cortex identified as V1 take the data from the optic nerve and send signals to a bean-sized area of neurons elsewhere in the visual cortex, identified as V4, where it is converted into colour. That is a slight simplification, in that, as Sacks put it, V4 'signals to and converses with a hundred other systems in the mind-brain' which interpret and apply meaning to colour. Mr I could only see – and remember – in black and white because his V4 cells had been damaged, leading Sacks to conclude that 'colours are not out there in the world but are constructed by the brain'.

Neuroscientist Bevil Conway compares the way our brains process colour to an iPhone: 'On the surface, it all seems incredibly simple – but there's a lot of complicated stuff going on underneath to make it feel simple.'

Every so often that complicated stuff confounds us. One famous example is the #dressgate Twitter storm about whether a photograph of a particular dress, posted by BuzzFeed in 2015, was white and gold or blue and black. In a single day, the post was viewed 28 million times, with two-thirds of people insisting the dress was white and gold. Intriguingly, a follow-up study of 1,400 respondents, published in the journal *Current Biology*

When a photo of this dress was posted on Twitter in 2015, two out of three people said it was white and gold. It was actually blue and black.

three months after the furore, found that 57 per cent thought it was blue and black – which the dress, made by British company Roman Originals, actually was.

There is no consensus as to why there should be such a discrepancy. Some have suggested that people's responses varied according to the device on which they viewed the image, or the light conditions in which they looked at it. Research suggested that early risers were more likely to describe the dress as white and gold, whereas 'night owls' thought it was black and blue. Another study showed that people with the most activity in the brain's frontal and parietal areas, which play a vital role in cognition, were more inclined to misperceive it as white and gold.

In 2015, when American neuroscientist Israel Abramov asked men and women to break down the hue of a colour and quantify how much red, yellow, green and blue it contained, he found that women were more likely to distinguish between subtle gradations of colour than men. This effect was particularly pronounced with hues that were mainly yellow and green. Abramov suggested that male understanding of colour may be inhibited by testosterone – men have more receptors for this hormone in the brain than women (especially in the parts that control vision) – but others argue that the cause is cultural.

A 1991 study by Jean Simpson and Arthur Tarrant suggested that women have a larger colour vocabulary than men – but found that age also plays a part, with older men using more elaborate colour terms than younger women. Other studies have concluded that women are more effective at matching colour chips to colour names and matching colours from memory.

For some people, colour is more than a visual phenomenon. At its most basic, synaesthesia (the term comes from the Greek words for 'perceive together') is a cognitive condition in which one sense triggers another. In his book *What Do You Care What Other People Think?* Nobel Prize-winning physicist Richard Feynman observes: 'When I see equations, I see the letters in colours. As I'm talking, I see vague pictures of … light tan j's, slightly violet-bluish n's and dark brown x's flying around and I wonder what the hell it must look like to the students.'

For some synaesthetes, like Atlanta pastry chef Taria Camerino, colour is a taste. In 2013, she told Audrey Carlsen on National Public Radio that she struggled to remember what things looked or sounded like, 'but I know what green tastes like'. Audrey Carlsen also interviewed British IT consultant and synaesthete James Wannerton, who tastes sounds, words and colours, and told her that her first name tasted strongly of tinned tomatoes.

When American psychologist Carol Crane hears guitar music, she feels something brush against her ankles.

We don't know conclusively what causes synaesthesia. British clinical psychologist Simon Baron Cohen argues it's a genetic condition, and that people with synaesthesia are born with more than the average number of neural connections. Some tests have suggested that synaesthetes have more myelin, a fatty sheath that surrounds neurons and helps signals to travel across the brain. It has been suggested that we are all born synaesthetic, but that we

Nobel-winning physicist Richard Feynman had synaesthesia, which allowed him to see equations in colours, as envisaged by Jim Ottaviani and illustrator Lelan Meyrick in their graphic novel on Feynman's life.

lose many neural connections in infancy to help our brains run more efficiently. Estimates vary as to how common synaesthesia might be, but it's possible that as many as one in 300 of us have some form of the condition.

In the preface to his 1943/44 work *Trois petites liturgies de la presence divine*, Olivier Messaien writes: 'The music is above all a music of colours. The "modes" that I use there are harmonic colours. Their juxtaposition and their superposition give: blues, reds, blues striped with red, mauves and greys spotted with orange, blues spiked with green and circled with gold, purple, hyacinth, violet, and the glittering of precious stones: rubies, sapphire, emerald, amethyst – all that in draperies, in waves, in swirling, in spirals, in interlaced movements. Each movement is assigned to one "kind" of [divine] presence … These inexpressible ideas are not expressed – they remain in the order of a dazzle of colours.'

In 1969, in their groundbreaking study *Basic Color Terms: Their Universality and Evolution*, American academics Brent Berlin and Paul Kay argued that eleven basic categories of colour were universal and that these basic colour terms always emerge in the same chronological order. After surveying more than a hundred languages, they concluded that the first two terms are always dark and light (usually interpreted as black and white); the third is red; the fourth yellow or green; the fifth is whichever of green or yellow is not already present; the sixth is blue; the seventh brown; and the eighth could be purple, pink, orange or grey.

In their definition, a basic colour term is not a compound (so red, not red-yellow), it's not qualified (blue, but not bluish), it's

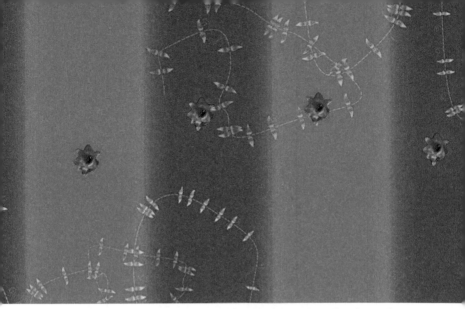

The pianist Håkon Austbø's colour visualisation of Messiaen's mode 33, from his article on the significance of Messiaen's colours in the Journal *Music & Practice*.

not a division of any other term (excluding, for example, crimson, which is a hue of red), it's not confined to a narrow range of objects (so no auburn, which is primarily used to describe hair colour), it's not the name for an object (so gold and silver are not basic), and it's not recently borrowed from another language.

The corollary of all this, Berlin and Kay contended, was that the more sophisticated languages have a greater number of colour terms than the more 'primitive' ones. Thus English has the full menu of eleven basic terms, but the speakers of Yéli Dnye in Papua New Guinea have to manage with just three colour words, having no labels for some 40 per cent of the visible spectrum.

Linguistic relativists argue that our colour vocabularies are a cultural construct. Is it valid, they ask, to talk of universal basic colour terms if the same term can mean radically different things in different societies? In the so-called 'grue' languages, the distinction between green and blue either does not exist (as is the case in Tzeltal, Lakota Sioux and Ossetian) or is significantly blurred (in Korean, the word *pureu-da* can mean blue, green or bluish green and in Vietnamese *xanh* can mean blue or green). To Russians, the distinction between light blue – *goluboy* – and dark blue – *sinly* – is as profound as that between blue and green in many other cultures. The Berinmo, a tribe of hunter-gatherers in Papua New Guinea, have five basic colour terms and, though they do not distinguish between blue and green, have two for shades of yellow. In the Filipino language of Hanunóo, *biru* can describe black, violet, indigo, dark green and dark grey.

The original Berlin and Kay thesis has been revised over the years to take some of these anomalies into account, but critics still argue that it's blighted by a Western cultural bias. In the Hanunó'o language, Geoffrey Sampson notes in his book *Educating Eve: The 'Language Instinct' Debate* (1997): 'The reference of colour terms are not even wholly determined by chromatic properties; it is partly determined by wetness or dryness … Perception of wetness or dryness can override the hue variable in determining the suitable colour word.' As American neuroscientist Bevil Conway says: 'People develop words for colour they talk about. In many societies, colour is always specific – it describes a fruit, a shade of textile or an animal's fur – and not an abstract quality.' Berlin and Kay's colour sequence may be applicable to the majority of codified languages, but that's not the same thing as a universal law.

'An object may be described of such a colour by one person and perhaps mistaken by another for quite a different tint,' the Edinburgh flower painter Patrick Syme complained. Keen to resolve such disputes, in 1814 he published Werner's *Nomenclature of Colours*, based on a taxonomy by German geologist

Abraham Gottlob Werner. Part nature nerd and part poet, Werner described colours with such precision that Charles Darwin took the book with him on his historic voyage on the HMS *Beagle* in 1839. In his way, Werner was a forerunner of Pantone. The 108 standard colours in the book can be modified by applying the terms 'pale, deep, dark, bright, dull and tinged with' as required. Each colour is represented with a small chip

YELLOWS.

No.	Names	Colours	ANIMAL	VEGETABLE	MINERAL
62	Sulphur Yellow.		Yellow Parts of large Dragon Fly.	Various Coloured Snap dragon.	Sulphur
63	Primrose Yellow.		Pale Canary Bird.	Wild Primrose	Pale coloured Sulphur.
64	Wax Yellow.		Larva of large Water Beetle.	Greenish Parts of Nonpareil Apple.	Semi Opal.
65	Lemon Yellow.		Large Wasp or Hornet.	Shrubby Goldilocks.	Yellow Orpiment.
66	Gamboge Yellow.		Wings of Goldfinch. Canary Bird.	Yellow Jasmine.	High coloured Sulphur.
67	Kings Yellow.		Head of Golden Pheasant.	Yellow Tulip. Cinque foil.	
68	Saffron Yellow.		Tail Coverts of Golden Pheasant.	Anthers of Saffron Crocus.	

Many of Werner's hues in his *Nomenclature of Colours* (1814) had functional names like lemon yellow; others were more esoteric – such as Gallstone Yellow.

and, where possible, likened to animals, vegetables or minerals. Orpiment Orange is, Werner suggests, the colour of belly of the warty newt, Skimmed-Milk White is equivalent to the 'White of Human Eyeballs' and Blackish Green is defined as 'Dark Streaks on Leaves of Cayenne Pepper'.

The comparisons were perhaps too poetic to fulfil Syme's goal, and in 1905 American artist and art teacher Albert Henry Munsell published *A Color Notation*, which specified colours based on three properties: basic colour, chroma (intensity) and value (lightness). A modified version of Munsell's system is still used to define the colour of teeth, soil and beer – and of skin and hair in forensic pathology. The Pantone guide, first published in 1963 by American printer Lawrence Herbert, is the most successful solution to Syme's problem, giving each colour a numerical value so that Pantone 17-1664 Poppy Red (the colour of Dorothy's red shoes in *The Wizard of Oz*) is the same hue wherever or however it is used.

❖ ❖ ❖

Pantone, which has built a lucrative business out of categorising the spectrum, has identified 1,867 colours for printing ink. Some of Pantone's colours have fancy Werner-esque names – notably Fallen Rock, Cloud Dancer and Grandma's Sweater – but its descriptions are significantly less evocative, especially when it hypes its Color of the Year. The 2019 choice, Living Coral, a peachy orange shade, was hailed as an 'animating and life-affirming golden hue' that welcomed and encouraged 'lighthearted activity'. Luxury paint maker Farrow & Ball is rather more Werner-like with its 'carefully curated' shades, offering colours such as Danish Lawn, Potted Shrimp, Elephant's Breath and Mouse's Back. In eighteenth-century China, even more imaginative labels were in use, such as 'camel lung' and 'dribbling spittle'. At around the same time, in France, there was a colour described as 'flea's belly', and another called 'Paris mud'.

SEEING RED

In the beginning, the colour was red. The Neanderthals, far from being the knuckle-dragging oafs of popular stereotype, were probably the first people to depart from monochrome. Around 64,000 years ago, in caves in what is now Spain, Neanderthals painted red horizontal and vertical lines on rocks, in the shape of a ladder. Forty thousand years ago, our ancestors painted their bodies with ground-up red clay. Between 35,000 and 15,000 years ago, people painted figures – mainly bison – on the ceiling of a cave in Altamira in northern Spain. The bison were engraved onto the ruddy rock and, primarily, painted in bichromatic red and black. The red colours were made from iron oxides such as haematite; the black, from charcoal.

For prehistoric painters, making black from charcoal was straightforward. Creating red was not. At first, they probably sourced colours from animal fat and blood, saliva, water and plant juice. When these pigments were found to fade quickly, they turned to minerals. Once they had mined the haematite, one of the most abundant minerals on Earth, they had to wash it, filter it and grind it to produce a fine red powder. After that, they had to discover which additives – animal fats or crushed bone

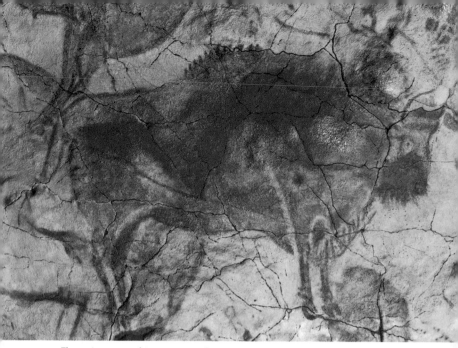

The extinct steppe bison, preserved in iron oxide at Altamira.

– would bind the paint, make it stick to the walls and spread across a large surface area.

They painted with their fingers at first, moving on to brushes of animal hair and rough crayons made out of lumps of pigment. Sometimes, they blew pigment onto the rock through tubes of reed or hollowed-out animal bones. Some of the Aboriginal artists who created the Bradshaw paintings, the awe-inspiring gallery of rock art in Kimberly, Western Australia, which is at least 17,500 years old, used feather quills to capture the fine details.

As Narayan Khandekar, the conservation expert who looks after the Forbes collection of pigments at Harvard, has observed: 'These artists chose colour. These people were hunting and surviving in inhospitable environments, but they were still taking the time to look out for colour and use it to tell the stories they wanted to tell. So, I think, to underestimate colour is to underestimate what people are.'

We still don't really know what stories these distant ancestors wanted to tell or why. Did their paintings have religious significance? Were they an early example of magical thinking, depicting the creatures they needed to kill for food? Were they trying to make sense of a violent, unpredictable world in which these animals were both prey and peril? Or were they designed to bind communities together, by inspiring collective action and reflection?

The allure of the colour itself – red, signifying raw meat, blood and fire – is easier to comprehend. When our ancestors learned to make fire – variously estimated to have occurred between one million and 100,000 years ago – and use it to provide light, heat and protection, it became so integral to their lives that in time it became an object of worship. Even though flames are typically yellow or orange, fire was invariably depicted as red – a colour later associated with Hephaestus, the Greek god of fire and metal; Agni, the Vedic Hindu fire god; and the unrelenting blazes of old school Christian hell.

As Bauhaus artist Josef Albers wrote: 'If one says red ... and there are fifty people listening, it can be expected that there will be fifty reds in their minds, and one can be sure that all those reds will be very different.' Red can spell danger (red light, red alert, red fire engines, red skies in the morning, a warning to shepherds and sailors and, more obscurely, the redcaps – the murderous goblins said to plague the Scottish Borderlands who reputedly painted their caps with human blood), love (hearts and roses on Valentine's Day), lust (the scarlet woman, red-light districts), anger and aggression (red mists, seeing red, red rag to a bull), bureaucracy (red tape and the red boxes of documents given to government ministers), embarrassment (as in the recurring *Private Eye* gag 'Red faces all round!'), socialism (of which more later), debt (going into the red), social status (the red carpet, first rolled out in the 1900s at New York railway station for first-class passengers), patriotism (it features on 155 national flags), mourning (in South Africa), guilt (red-handed), fine weather (red skies at night), Wales (the red dragon), fast cars (especially Ferrari), top sports teams (Manchester United, Liverpool, Chicago Bulls) and the Devil (described in the Book of Revelation as a red dragon with seven heads and ten horns).

'The blind live in an undefined world from which certain colours emerge. As for red, it has vanished completely. But I hope some day – I am following a treatment – to be able to see that great colour, that colour which shines in poetry, and which has so many beautiful names in many languages. Think of scharlach *in German,* scarlet *in English,* escarlata *in Spanish,* écarlate *in French.'*

Jorge Luis Borges, who went blind at the age of 55

In Gerry Anderson's TV show *Captain Scarlet and the Mysterons*, an organisation called Spectrum tries to defend the Earth against angry, invisible aliens. Calling this august body Spectrum was a political statement by Anderson, whose previous attempts to integrate ethnic minorities into his Supermarionation puppet dramas had been stymied by fears that they would not be televised in the American South.

By 1967, even US TV network execs recognised the need for change. The new series' cast included Trinidadian Lieutenant Green and Chinese-Japanese woman pilot Harmony Angel. The indomitable Captain Scarlet, rendered indestructible through accidental exposure to radiation, fought alongside Captains Blue, Brown, Grey, Magenta and Ochre, Lieutenant Green and Doctor Fawn. Anderson was seriously peeved when, after the series was rerun in the 1970s, some critics argued that it had racist undertones, on the basis that Spectrum was led by the benevolent Colonel White and the Mysterons' most dangerous agent was

Captain Scarlet – a colourful futuristic drama, originally revealed to British TV viewers in black and white.

Captain Black. (To add to the colourful confusion, White and Black were both voiced by South African actor Donald Gray.)

Anderson's dark fantasy has been interpreted as a metaphor for the Cold War (which would make Captain Black politically red) and as a Christian allegory. In this latter reading, White stands for God, Black for the Devil, Scarlet for Jesus and his holy blood. The oft-mentioned slogan 'Spectrum is green' reflects the belief, according to Cy Grant, the actor who voiced Lieutenant Green (and was actually Guyanese), that green is the 'healing colour of nature'.

Ironically, this colourful futuristic drama, premiered on 29 September 1967 on ATV, the commercial TV franchise for the Midlands, in black and white. Only later, in 1970s repeats, did British viewers see the vivid spectrum of *Captain Scarlet and the Mysterons* in all its Supermarionated glory.

'One of the Englishmen we had shot down enquired after the Red Aeroplane. In his squadron, there was a rumour that the Red Machine was occupied by a girl, a kind of Jeanne d'Arc. He was extremely surprised when I told him the girl was standing in front of him. He was convinced that only a girl could sit in the extravagantly painted machine.'

Manfred von RIchtoften, The Red Battle Flyer (1918)

German World War I flying ace Manfred von Richthofen, aka the Red Baron, flew a red Fokker aircraft from December 1916 until he was shot down and killed near Amiens on 21 April 1918. Von Richthofen may be the most famous historical 'Red', but there are two Red Duchesses (Katharine Stewart-Murray, a Scottish Unionist who supported the Republicans against Franco

The Red Baron and his red Fokker.

in the Spanish Civil War, and Luisa Isabel Álvarez de Toledo, a left-wing Spanish aristocrat), a Red Priest (Antonio Vivaldi, so known because of his hair colour, much like Norse adventurer Erik the Red), a Red Prince (Wilhelm von Habsburg, who aspired to become the socialist king of Ukraine after the First World War) and a Red Dean (Hewlett Johnson, the Dean of Canterbury whose unswerving devotion to the Soviet Union, and Stalinism, earned him the Order of the Red Banner of Labour).

❖ ❖ ❖

Psychologists agree that red influences the way we think, feel and behave, but they don't quite agree on how that influence works. Some studies suggest that exposure to red lowers our scores in IQ tests and makes it harder for us to think analytically. Others say that it helps us to perform simple tasks and do clerical work. In a 2009 paper on colour and cognitive performance, academics Ravi Mehta and Rui Zhu analysed six studies which collectively suggested that red can help us stay vigilant, remember things and read proofs accurately, but inhibits creative thinking. They argued: 'Red, because of its association with dangers and mistakes, should activate an avoidance motivation, which has been shown to make people more vigilant and risk-averse.'

When we get angry, oxygenated blood rushes into our facial veins, creating a red flush. It's no surprise, then, that in many cultures there is a close association between red and anger. In Britain, the connection goes back at least four hundred years. The *Oxford English Dictionary* references a line in Thomas Heywood's epic poem *Troia Britanica* (1609): 'But his red wrath King Nestor did restrain.' In Shakespeare's *King John*, probably written between 1594 and 1596, the suspicion that the king ordered the murder of his nephew Arthur, Duke of Brittany, so inflames England's barons that they vow 'with eyes as red as new-kindled fire' to find the victim's grave.

The idea of 'seeing red' when one is furious is found in many other languages – German has the verb *rotsehen* ('to see red'), and in Swahili someone who becomes very angry is likewise 'seeing red'. In Spanish there's the phrase *verlo todo rojo* ('to see everything red'), but the link to a bull being infuriated by a cape-waving matador is deceptively obvious – bulls can't actually see that colour.

People do report that they have literally 'seen red' when angry, but this seems to be an exceptional occurrence. In these

Riley sees red in Pixar's animation *Inside Out*.

instances of 'red mist', it's likely that extreme rage causes the blood vessels in the eye to swell, in effect putting a red filter over the photoreceptors in the retina. If 'seeing red' were a common physiological reaction, you would expect the association with anger to be more universal than it is.

One species renowned for seeing red is Britain's favourite bird, the robin (*Erithacus rubecula*). A heart-warming presence on Christmas cards since the nineteenth century, revered by William Blake ('A robin redbreast in a cage/Puts all nature in a rage', he declared in 'Auguries of Innocence'), this popular bird is, in fact, as *New Scientist* magazine succinctly put it, 'a red-breasted thug'.

Robins fluff up their red breasts to defend their domain. If that doesn't work, they fight intruders to the death. Ornithologists estimate that one in ten adult robins die from fractured

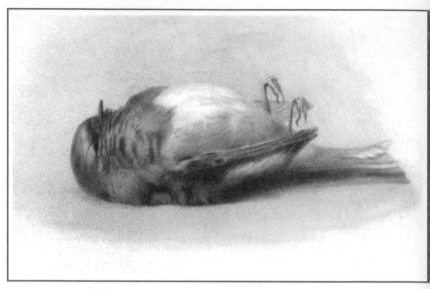

May yours be a Joyful Christmas.

Britain's favourite bird – in reality a red-breasted thug – celebrated in recumbent form on this (genuine) Victorian Christmas card.

skulls inflicted during such disputes. And it is the red that gets their dander up. Experiments by British scientist David Lack and others have shown that they are so infuriated by the colour that, in the absence of live red-breasted opponents, they will attack dead robins, their own reflection, plumes of red feathers and, very rarely, a red-bearded human.

In the folklore of Brittany and Wales, Christ's blood is said to have given the robin its red breast because the bird had swooped to take a thorn out of Jesus's crown of suffering.

❖ ❖ ❖

We tend to think that Rudolph, the robin's chief Christmas rival, has a red nose as a long-standing festive tradition. The famous Christmas poem 'A visit from St Nicholas' (better known as ''Twas The Night Before Christmas'), written in 1823 by American theologian Clement Clarke Moore, celebrates the eight reindeer that pull Santa's sleigh, but does not mention Rudolph. It was in 1939 that the American department store Montgomery Ward published Robert May's festive promotional story 'Rudolph the Red-Nosed Reindeer', which inspired Johnny Marks to write the song that became a big hit for Gene Autry ten years later. The red nose was obviously alliterative but also reflected the green and red colour scheme that then characterised Christmas. Rudolph's red nose is still shining more than eighty years later.

Less than 2 per cent of people on this planet are red-headed, making red the rarest of human hair colours. In Scotland, however, around 12 per cent of the population are natural red-heads, while in Ireland the figure is about 10 percent. Why should this be? In the great majority of cases, red-haired people have a variant of the melanocortin 1 receptor on chromosome 16, which leads to higher than average concentrations of the pigment called pheomelanin, and much lower concentrations of the dark pigment known as eumelanin. High levels of pheomelanin don't just produce red hair – they also produce paler skin. Lighter skin pigmentation makes it easier for the body to generate the necessary levels of vitamin D in climates that have a low level of sunlight. Hence, it's been argued, the greater number of pale-skinned russet-haired humans in the cooler zones. In Africa, by contrast, red hair is at an evolutionary disadvantage because pale skin is harmed by strong sunlight.

The root of all the superstitions, legends and stereotypes sur-rounding redheads is surely that they are rare and, therefore, conspicuous. Roman historians, for example, were intrigued by the wild redheads they encountered as the empire expanded northwards into Gaul, Germany and Britain. In various parts of the world, redheads acquired a reputation for being promiscuous, hot-tempered, hard-headed, malicious or plain evil. *Brewer's Dictionary of Phrase and Fable* records that 'the fat of a dead red-haired person used to be in demand as an ingredient for poison.'

The misinformation that swirls around colour is epitomised by the folklore surrounding red cars. They are variously alleged to be driven by more aggressive motorists, accumulate an inordinate number of speeding tickets and cause a disproportionate number of accidents, all of which explains why, as the internet assures us, red cars are banned in Brazil. Which they aren't.

The legend that red cars cost more to insure because they are more accident prone still endures, as does the theory that the sight of red cars stimulates an aggressive reaction from other drivers, who literally and emotionally 'see red'. An exhaustive study by America's AAA Foundation for Traffic Safety concluded in 2004: 'The bottom line is that there is presently no scientific evidence supporting the selection of one particular vehicle colour as the unambiguous best choice for safety.'

'Ask a child to draw a car and, certainly, he will draw it red.'
A remark reputedly made by Enzo Ferrari

Ferrari's luxurious sports cars and iconic Formula One machines have made the company one of the world's most charismatic brands, and colour has been integral to its success. At one point in the early 1990s, 85 per cent of Ferraris were painted red. Even today, around four out of ten are that colour.

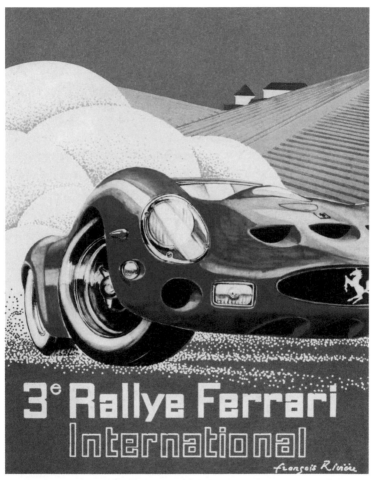

The French countryside is left behind in a cloud of smoke in this 1960s poster for the Ferrari rally.

The notion of a definitive 'Ferrari red' is misleading, because the cars are available in a multiplicity of different shades, but the colour that comes most readily to mind when we think of Ferrari is Rosso Corsa, or 'Racing Red'. In 1907, the governing body of the nascent sport of motor racing decided that countries must compete under specific colours. As the nation that practically invented motor racing, France probably had first choice, opting for the blue in its tricolour. Italy chose red, which may, as Ferrari's biographer Richard Williams suggests, have been a homage to Garibaldi's red shirts. It may also have been inspired by the fact that Prince Scipione Borghese won that year's Peking to Paris rally in a red 40hp Itala car, made in Turin.

On 6 September 1908, the ten-year-old Enzo Ferrari watched his first Grand Prix with his father and brother, in Bologna. The race was won by Turin's Felice Nazzaro, who averaged 120kph over the 50-kilometre course in a 'Racing Red' Fiat.

Enzo was so entranced by the spectacle he decided to become a racing driver, irritating his father Dino, who wanted him to enter the family metal fabrication business. Ferrari's love of red probably reflects his early memories of those races, although it is worth noting that the family car, in which his father drove away to join the Red Cross in 1915, was a red four-cylinder Diatto Torpedo. A year later, having transferred to the Italian Air Force, Dino died of bronchitis and pneumonia.

When Ferrari began competing in Formula One in 1947, the old regulations on colour still applied, so Racing Red became the marque's trademark. Yet there was a point, near the end of the 1964 Grand Prix season, when Ferrari nearly gave up the colour. The sport's governing body, the Fédération Internationale de l'Automobile, had refused to approve the company's new car, the 250 LM, after complaints by rival teams that it was cheating (by not making the required minimum number of vehicles). When Italy's motor-racing organisation failed to support him, an enraged Enzo swore his team would never race in red again. For the final two races of the season, under the nom de plume NART, the team competed in blue and white. The change of name and colour made no difference to John Surtees, who won the last Grand Prix in Mexico – and with it the world championship. As a compromise was later reached, Surtees remains the only Ferrari driver to win the Formula One title in a car that wasn't painted red.

On 28 August 2018, an event of apocalyptic import was announced on YouTube. The clip – posted by delighted members of a fundamentalist Jewish group called the Temple Institute – revealed that a 'perfectly red heifer' had been born in Israel.

Founded by Rabbi Yisrael Ariel in 1987, the Temple Institute is devoted to the creation of a successor to the first two Holy Temples of Jerusalem, which were destroyed by the Babylonians and the Romans. According to the Book of Numbers, the ashes of a perfect red heifer mixed with water are a necessary element for purifying the people of Israel at the Temple. Tradition has it that only nine such heifers were sacrificed in the first two temples. The tenth will be used by the Messiah himself, whose

The Temple Institute's red heifers – the end of the world as we know it?

arrival will bring about the rebuilding of the Temple, which in turn will precede the end of the world.

Hence the excitement at the Temple Institute, which has been trying to produce such a rare specimen for a long time, and in 2015 launched a $125,000 crowdfunding appeal to finance the import of Red Angus cow embryos. The Institute has recreated many artefacts for use in the Third Temple, but has not publicly pronounced on the progress of its unblemished heifer since the autumn of 2018. Nor has it explained how it thinks the Temple might rise on the spot that's currently occupied by the Dome of the Rock, one of Islam's holiest sites. If the worst should come to pass, the *New York Times* has already provided the perfect headline: 'Apocalypse cow'.

In 1728, the visual impact of red – and its association with danger – persuaded British inventor Richard Newsham to use this colour to distinguish his fire engines, two of which were later sold to New York. Red became synonymous with firefighting in large American cities, but in many smaller communities the engines were lime green or yellow. In the UK, fire engines likewise became predominantly red during the nineteenth century, but by the 1960s the iconic red fire engine was undergoing something of a re-evaluation. Albert Leese, the chief fire officer for Coventry between 1960 and 1974, ran some tests at the local Lanchester College, with experts from Dulux paint company. They concluded that engines would be more visible if they were somewhere on the spectrum between lime green and yellow. Because the repainted engines were involved in fewer accidents, Leese urged the Home Office to switch to the new colours nationally. The Home Office replied: 'Fire engines have always been red.'

Leese's conclusions were supported by further research in America but were challenged by a 2009 study for the US Fire Administration, which suggested that lime-green engines may not be as visible as tests predicted. Retroreflective striping, this study concluded, would be more effective.

Sotheby's director Philip Hook, in his book *Breakfast at Sotheby's*, reports that the value of Piet Mondrian's paintings is directly proportional to the amount of red in them.

Mondrian's *Composition II in Red, Blue, and Yellow*, 1930.

❖ ❖ ❖

In the summer of 2014 in China, the popularity of American TV shows such as *True Blood* and *The Vampire Diaries* made the 'world's first blood substitute beverage' a trendy drink for teenagers. Packaged in an authentic-looking blood bag and – the makers proclaimed – possessing the 'same texture, colour and nutrients as blood', this drink, and others like it, became so popular that the country's Food and Drug Administration banned them because they violated 'social integrity and moral principles and fail to conform to related national laws.'

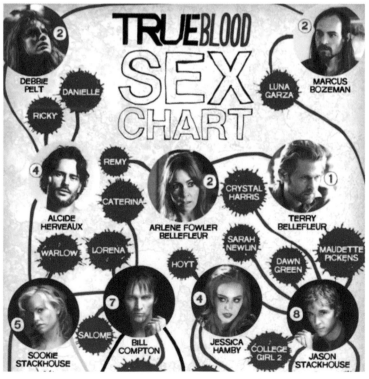

True Blood revamped vampire shows with a lot of sex, mapped here by MTV.

Chinese teenagers were only pretending to break one of the oldest, strongest and most widespread social taboos, but members of the so-called *sang* community (from the French word for blood) actually do drink human blood, to cure their ills or boost their psychic energy.

Healthy blood is always red, but it comes in different shades. Arterial blood, freshly oxygenated by the lungs, is brighter than the blood in our veins. The distinction was not lost on John Keats, who trained as an apothecary and surgeon before focusing on poetry. On 3 February 1820, he coughed a small spot of blood onto a bedsheet at Wentworth Place (now Keats House) in Hampstead. Inspecting the blood by candlelight, he remarked: 'I know the colour of that blood, it is arterial blood. That drop of blood is my death warrant.' His diagnosis was tragically accurate: he died of tuberculosis, on 20 February 1821, in Rome, at the age of 25.

The most famous blood in Western culture is that shed by Jesus Christ on the cross. In Christian theology, his blood symbolises a new covenant which promises the remission of sins for those who believe in God. In the gospels of Luke, Mark and Matthew, Christ tells his disciples to eat bread and drink wine in memory of him. In the subsequent Christian rite of the Eucharist, the faithful drink from a cup of wine which, depending on which variety of Christianity is involved, is either miraculously transformed into Jesus's blood or signifies his spiritual presence.

Christ's agony is symbolised by the red cross, worn by the Knights Templar and featured on the flags of England, Georgia and the Canadian provinces of Alberta, Manitoba and Ontario.

The blood spilled by Christ was caught, the thirteenth-century French poet Robert de Boron claimed, by Joseph of Arimathea in a cup or chalice known as the Holy Grail, a vessel that de Boron linked to Arthurian legend. Other writers, notably Wolfram von Eschenbach in his romance poem *Parzival*, identified the Templars as guardians of the Holy Grail.

Seven centuries later, blood, grail and Templars were woven into a ludicrous conspiracy theory suggesting that a French draughtsman named Pierre Plantard (who died in 2000) was descended from Christ, who supposedly had fathered a child with Mary Magdalene. In Dan Brown's mega-seller *The Da Vinci Code*, Plantard metamorphoses into sexy French cryptologist Sophie Neveu, the 'last living descendant of Jesus Christ'.

In 2007, Italian football club Inter Milan were criticised for wearing white shirts with a red cross in a match against Turkish champions Fenerbahce. In an official complaint to FIFA, Turkish lawyer Bariş Kaşka lamented: 'That cross brings only one thing to mind – the Templar knights. While I was watching the game, I felt profound grief in my soul.' Kaşka might not have realised that a red cross on a white background was the symbol of Ambrose, the fourth-century bishop who is Milan's patron saint.

Before the First World War, the Flanders region of Belgium and northern France was not renowned for its red poppies. Yet by early 1915, they had become so common that soldiers mentioned them in their letters home. Constant bombardment had disturbed the soil in Flanders, bringing the seeds to the surface. The nitrogen in the explosives – and the lime from shattered buildings – and

Inter Milan blithely impervious to Crusader connotations in their 'Templar' shirts.

the blood and bones of the slaughtered men and animals then fertilised them. Every fallen shell, every fallen soldier, made the ground more suitable for poppies. The flowers were celebrated in John McCrae's poem 'In Flanders Field', and less famously by British lieutenant-colonel and poet W. Campbell Galbraith: 'The slender waving blossoms red/Mid yellow fields forlorn/A glory on the scene they shed/Red Poppies in the Corn.' Such lines reinforced the poppy's resonance as a poignant symbol of this unimaginable carnage.

Our imperfect knowledge of ancient history makes it hard to define when red first became a martial colour. Historian Tom Holland says that Lycurgus (800–730 BCE), the leader who reformed Sparta into a military superpower, personally decreed that his army must wear 'brilliant cloaks, dyed the colour of fresh blood'. Greek historian Xenophon mentions the 'wall of bronze [shields] and scarlet' with which the Spartans went into battle.

Scarlet uniforms could be interpreted as a warning to opponents, implying that the Spartan troops were already steeped in the blood of vanquished enemies. In 1958, American scientist Robert Girard documented the effects of red, white and blue lights on the visual cortex and nervous system and found that red increased the viewer's blood pressure, respiration and anxiety levels.

The first British troops known to have worn red uniforms were those fighting for Queen Elizabeth I against rebels in Ireland – in 1561, they lost the 'Battle of the Red Cassocks' (the Irish word *casóg* can also mean coat, cloak or uniform.)

In 1621, in his history of the Tudor conquest of Ireland, Irish soldier Philip O'Sullivan Beare celebrated the rebels' victory, writing: 'Among those who fell in battle were four hundred soldiers lately brought from Britain and clad in the red livery of the viceroy.' Oliver Cromwell's New Model Army standardised a red infantry uniform, wearing it in battle near Dunkirk in 1658, during the Franco-Spanish War. (At the start of the English Civil War, both Royalists and Parliamentarians wore red.) By 1740, *The Craftsman* magazine was describing the British infantryman as 'Thomas Lobster'.

Last stand of the red coats at Gennis.

The last time the British wore red coats in battle en masse was on 30 December 1885 at Gennis, Sudan, against Sudanese-Islamic rebels. In 1902, khaki became the British Army's battle dress, although red is still worn by the Foot Guards and Life Guards at ceremonies such as Trooping the Colour, the celebration of the monarch's official birthday. The descendants of the British imperial red coat are also still worn ceremonially by armies in former colonies such as Australia, Fiji, Ghana, Jamaica, New Zealand, Singapore, Sri Lanka, and it inspired the red serge jacket worn by the Royal Canadian Mountain Police.

Britain's iconic red postboxes, introduced in 1874 because people kept bumping into the existing green 'pillar boxes', are still much in use today in Australia, Belize, Canada, Cyprus, India, New Zealand, Pakistan and South Africa. Red telephone boxes, which first appeared in Britain in 1924, are still found in Antigua, Australia, Barbados, Cyprus and Malta.

British imperial red's colonial significance was not lost on the Chinese government when it regained control of Hong Kong in 1997. The postboxes were quickly repainted green – the same colour as in mainland China – but a few retained their royal insignia until 2015. Furious protests by local conservationists struck the Chinese government, in the words of one retired senior Beijing official, as proof of Hong Kong's 'failure of decolonisation'.

The famous *bonnet rouge* (red cap) worn by French revolutionaries as a symbol of solidarity and defiance was a derivative of the 'Phrygian cap', a close-fitting conical hat, made out of soft felt or wool, originally worn by the Phrygians, who ruled much of what is now Turkey from the twelfth century BCE to the seventh century BCE. The association between liberty and this type of headwear can be traced back to ancient Rome, where a cap called a pileus – the symbol of Libertas, the goddess of liberty – played a pivotal part in the ceremony to mark the emancipation of a slave. The synergy between hat and freedom was reaffirmed after Nero's death in 68 CE when, the historian Suetonius records, 'such was the public rejoicing that the people put on liberty caps and ran all over the city'.

Red caps as a symbol of liberation first appear in French history in 1675, with an uprising of Breton peasants against Louis XIV. Their adoption of red caps – although they may also have worn blue ones – set the precedent for the *bonnet rouge* worn by the revolutionaries overthrowing France's *ancien regime*. The Jacobins were probably also influenced by the Americans, who put the red cap to symbolic use during the War of Independence which, with a lot of help from the French, they won in 1783. The *Révolutions de Paris* newspaper lauded the *bonnet rouge* as the 'emblem of emancipation from all servitude and the rallying sign for all enemies of despotism'. Under the Directorate, from 1795 to 1799, it was worn less often, and it was banned by the restored Bourbon monarchy. The ban did not last long, however, and nowadays the figure of

Revolutionary road: red-capped *sans culottes*.

Marianne – emblem of the French Republic – is usually depicted in a Phrygian cap.

On 17 July 1791, a crowd of 50,000 Parisians gathered in the Champ de Mars, demanding the dethronement of King Louis XVI. When the rally threatened to turn riotous, Bailly, the mayor of Paris, had a red flag raised, to warn the crowds to disperse. Before the throng could react, the National Guard, commanded by the Marquis de Lafayette, opened fire, killing around fifty people. The dead were remembered as 'martyrs of the revolution' and the red flag, emblematic of their blood, was adopted by the Jacobins as the banner of the revolution.

Six years later, in May 1797, with revolutionary France at war with 'perfidious Albion', overworked, underpaid British sailors hoisted the 'bloody flag' near Nore, in the Thames Estuary. The mutineers seized enough ships to partially blockade London and demanded that King George III make peace with France. This is the first time that a red flag is known to have been flown in an industrial dispute. As with the Jacobins, the red came to symbolise the blood of the martyrs: 29 rebellious sailors were hanged.

The first leader to wave the red flag as a symbol of liberty might not have been Robespierre, Karl Marx or Lenin. In 1347, the Italian notary Nicola Gabrini, better known as Cola di Rienzo, launched a coup to become 'tribune of Rome', a triumph secured by a procession in which, Luigi Barzini writes, 'the first flag dedicated to liberty was red'. Historians still debate whether Cola di

Rienzo was a populist, proto-Fascist or a liberal hero, but he is, as far as we know, the first person to have used the red flag as the people's standard.

'Favourite colour: red', said Karl Marx, in 1865, replying to a set list of questions that what we now know as the Proust Questionnaire (as regularly featured on the back page of *Vanity Fair* magazine). On 19 May 1849, after the Prussian authorities had

Marx sets his favourite colour in print.

closed Marx's newspaper *Neue Rheinische Zeitung*, its indignant editor-in-chief printed the last issue in red ink.

After the Franco-Prussian War, an alliance of radicals, socialists and anarchists – known as the Commune – swept to power in Paris's municipal elections in March 1871. Though the Communards commanded 380,000 battle-hardened national guards, their socialist regime was not allowed to last long. During 'Bloody Week' – 21–28 May 1871 – the French army retook Paris, street by street, probably killing 20,000 people. On 25 May, Louis Charles Delescluze, the Commune's last leader, donned his red sash of office, climbed to the top of the nearest barricade and was shot to death.

As Marx gleefully noted in his pamphlet *The Civil War in France*, 'the old world writhed in convulsions of rage at the sight of the Red Flag, the symbol of the Republic of Labour, floating over the Hotel de Ville.' The tragic splendour of the Paris Commune helped to establish red as the defining colour of the left. The *New York Times*, reporting on the Commune's third anniversary celebrations in Paris, headlined its story: 'An Assemblage Where Every Shade of Red Could Be Seen'. The article noted: 'Children wore red sashes, red dresses and red hats, and men adorned themselves with satin neckties of the approved roseate hue. Many of the young girls present, who took part in the ball afterward, had seized the opportunity, in their Communistic sympathy, to spread the patriotic colour on their faces.'

In the ferment of 1917, the red flag was flown by Russian revolutionaries of all types (they could make a narrow red flag simply by tearing the blue and white stripes off the traditional Russian flag), and when the Bolsheviks seized power they adopted it as their own. The red Soviet flag then defined the palette for socialist and communist republics. In Mongolia, the revolutionary government changed the capital city's name from Nïïslel Khüree ('Capital Camp') to Ulanbaataar ('Red Hero'). In China, Mao built his cult with the so-called *Little Red Book* (formally entitled *Quotations from Chairman Mao Tse-tung*), a collection of 427 aphorisms positioning him as a Marxist Confucius. In 2019, president Xi Jinping launched his own little red smartphone app, called *Study the Great Nation*, extolling his achievements. His app quickly acquired more than one hundred million registered users, possibly because people knew they would be tested on their knowledge of the app at work or school and shamed if their scores were too low.

❖ ❖ ❖

In Russia, *krasny* (the word for 'red') has the same root as *krasivy* (an old word for 'beautiful'). Moscow's glorious Red Square is named for its beauty, not its revolutionary significance.

❖ ❖ ❖

Red is lucky in China, being traditionally worn by brides but never at funerals. The custom of gifting money in red envelopes ('hongbao') dates back to the Qin dynasty (221–206 BCE) when the elderly would thread coins together with red string.

❖ ❖ ❖

Sensationalising the 'horrors' of the Paris Commune was the original manifestation of the institutionalised antipathy that led to 'Red Scares' in America after both world wars. The witch hunt of the 1950s – instigated by the House Committee of Unamerican Activities and spearheaded by Republican Senator Joseph McCarthy – is infamous. The first 'Red Peril' panic, which began in 1919, less so. The prime movers included Republican Attorney General A. Mitchell Palmer, Democratic Senator Lee Slater Overman and the young J. Edgar Hoover, head of the General Intelligence Division.

The Catholic Catechetical Guild of Minneapolis published these two Red Scare comics in 1947 (right) and 1960 (left).

Hundreds of suspected Communists were deported. Public display of the red flag was prohibited in many American states, a move declared unconstitutional by the Supreme Court in 1931. In this context, Oklahoma's official standard – a red flag, with a white star on it – became problematic. In 1925, the Sooner State adopted a new blue flag, even though the name 'Oklahoma' derives from the Choctaw words meaning 'red people'.

The fervour with which General Alfredo Stroessner's regime repressed Paraguay's 'Communist agitators' – a term applied to anyone he did not like or disagreed with – did not prevent his Colorado (literally 'coloured red' in Spanish) Party adopting a red flag with a white star in the top left corner as its official standard, and handing out red banners and red armbands at rallies. In Stroessner's Paraguay, red was a right-wing colour, although, confusingly, it could be used as a pejorative term against 'left-wing' opponents such as the outspoken Bishop Melanio Medina (the 'Red Bishop'), while the liberal opposition rallied behind blue.

In coverage of US elections, red states are those that lean to the Republicans, and blue states are those that side with the Democrats. And yet, in 1976, in the first colour-coded electoral map on TV, NBC marked the states backing Republican Gerald Ford in blue, and the states voting for Democrat Jimmy Carter in red. Republicans were traditionally blue, the colour worn by the Union Army, which fought under Republican President Abraham Lincoln in the Civil War. In 1984, CBS decided to use red for Republican and blue for Democrat. It wasn't until 2000, however, that this was adopted across the major news media. States

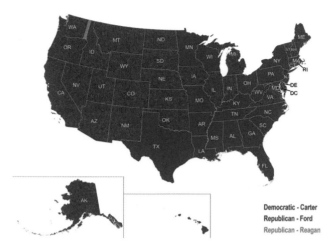

Democratic - Carter
Republican - Ford
Republican - Reagan

America's first colour election in 1976: Middle America is solidly Republican blue, but Carter sweeps the US for the red Democrats. The grey strip represents a vote for Ronald Reagan by a 'faithless elector' from Washington state.

that swing between the two main parties at elections, especially presidential ones, are known as 'purple states', a classification that includes Colorado, Florida, Iowa, Michigan, Minnesota, Ohio, Nevada, New Hampshire, North Carolina, Pennsylvania, Virginia and Wisconsin.

Wearing shades of red makes women more attractive to men, according to a 2010 study led by Andrew Elliot, professor of psychology at the University of Rochester. In the US and Europe, they showed men and women photographs of women on different coloured backgrounds and asked how attractive they found them. Men were more likely to find women on a red background attractive, whereas women were not. Men also preferred women photographed wearing a red shirt to those dressed in blue. Keen to explore how universal the 'red effect' is, Elliot led similar experiments in a small community in Burkina Faso in 2012. Even in such a different culture, the 'red effect' worked, prompting the speculation that 'Red may operate as something of a lingua franca in the human mating game.'

Do women take advantage of red's sexual allure? A study by Daniela Niesta Kayser at Germany's University of Potsdam in 2016 suggests they might. She invited women to her lab for an experiment, emailing them a photo of the researcher in charge. Half the women were emailed a photograph of a man who had been previously rated 6.6 (1 of 9) for attractiveness by twenty women. The other half received a photo of a man rated as 3.9. When the volunteers arrived, 57 per cent of those who were expecting to meet the attractive researcher wore red, pink or scarlet, compared to 16 per cent of those expecting to meet the unattractive man.

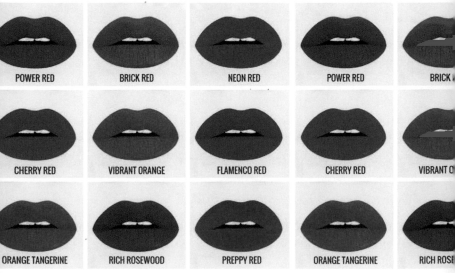

Lipstick shades of red from Aromi.

A 2012 study by sociologists at the Université de Bretagne-Sud found that waitresses who wear red lipstick get bigger tips – from men, that is.

The idea of red lipstick as a symbol of sexual arousal came to the fore in ancient Greece, where prostitutes were punished for not painting their lips red. The sensual aspect of red lips was reputedly underlined by Poppaea Sabina (30–65 CE), Nero's second wife. A hedonistic beauty who pioneered the celebrity hairstyle – thousands of Roman women dyed their hair amber to look like her – she was so fond of purplish and red lipstick she employed attendants whose only job was to paint her lips.

Wealthy Roman women mixed such ingredients as ochre, iron ore, lead, vermilion and fucus, a mercuric plant, to create a red lip paint that looked great, stank and was often toxic. Poorer women, who used a dye derived from red wine, may not have pouted as gorgeously but possibly lived longer.

Moviemakers have long interpreted red as a metaphor for sex and sin. In 1938, in the black-and-white antebellum melodrama *Jezebel*, Bette Davis's strong-minded heroine ignores her aunt's advice ('You know you can't wear red to the Olympus Ball') and is dumped by her fiancé for her 'sins'. Only a year later, with a much bigger budget, the perfidy of Scarlett O'Hara (Vivien Leigh) was displayed in vibrant Technicolor in *Gone with the Wind*. After her husband catches Scarlett flirting (her name is a bit of a giveaway), he orders her to wear a red dress to a party.

The equation between red dresses and sex became especially blatant in the 1970s and 1980s, with Yves Robert's comedy

Pardon Mon Affaire and the Hollywood remake, *The Woman in Red*, starring Gene Wilder. In both movies, a woman in a red dress re-enacts the famous scene in *The Seven Year Itch* in which a strong breeze from a subway grating lifts Marilyn Monroe's dress. In Hollywood's remake of Robert's comedy, Kelly LeBrock cavorts like a pole dancer in search of a pole. It's a crass scene, but the 'red effect' did wonders for her career.

Women in red were much on the mind of whoever wrote the Book of Revelation, between 60 and 95 CE. The elusive author probably wasn't John the Evangelist, but they certainly believed that red stood for damnation. The Whore of Babylon is sitting 'upon a scarlet beast', 'arrayed in purple and scarlet' and drunk on the 'blood of the saints' and 'the blood of the martyrs of Jesus'. In the West, scarlet has, at different times, been culturally synonymous with prostitution, Satanism and adultery – the offence for which Nathaniel Hawthorne's protagonist Hester Prynne is ordered to wear the scarlet letter 'A' by townspeople who seem to have forgotten the words of the Lord, as recorded in the Book of Isaiah: 'Though your sins are like scarlet, they shall be as white as snow.' Possibly due to the lack of any biblical precedent, there is, as yet, no colour to identify men of easy virtue.

In 1993, fashion designer Christian Louboutin was trying to perfect a pair of red-soled shoes. 'The two-dimensional sketches were so powerful, but the three-dimensional object was somehow lacking energy,' he recalled. 'Trying different things to liven up the design, I grabbed my assistant's red nail polish and painted the sole. I knew instantly this would be a success.' Louboutin later trademarked his red soles. The Pantone reference is 18-1663 TPX.

The ruby slippers worn by Judy Garland's character Dorothy in *The Wizard of Oz* (1939) were created by Adrian, MGM's chief costume designer, to make the most of the gloriously saturated hues of Technicolor. (In Frank L. Baum's novel, Dorothy wore silver footwear.) Adrian did his utmost to make sure that Dorothy's ruby red slippers sparkled on screen. The white silk pumps were dyed red and covered in a red silk fabric sewn with rows of more than two thousand dark red sequins. (They had to be dark red because, under the intense lights Technicolor used at the time, anything lighter would have looked orange.) In a lavish final touch, he added butterfly-shaped red leather bows decorated with red rhinestones, bugle beads and costume jewels.

In *The Wizard of Oz*, Dorothy's ruby slippers are a force for good, but for Karen, the protagonist of Hans Christian Andersen's morbid fairy tale *The Red Shoes*, colourful footwear signifies unbecoming vanity. A surprisingly ruthless angel decides that, as a punishment for such immodesty, Karen must dance to her death. Nothing – not even amputation – will stop her feet, and off they dance into the forest.

In the 1948 film of Andersen's story by Michael Powell and Emeric Pressburger, the red shoes are more ambiguous. We are unsure whether they murder the ballet dancer (played by Moira Shearer) who is wearing them as she jumps in front of a train or signify her determination to die for her art. Powell certainly believed art was worth dying for, which may explain why the heroine is warned by her deranged yet brilliant ballet director: 'A dancer who relies on the dubious comforts of human love can never be a great dancer.' That line resonated with one of the movie's fans, Kate Bush, whose seventh album, *The Red Shoes* was released in 1993. Mourning her mother, and the end of a long-standing romantic relationship, Bush didn't die for her art but she did exhaust herself writing, singing and producing the album. She did not release another album for twelve years.

Unilever exploited our innate tendency to assign meanings to colour when it launched Stripe (Signal in the UK), a white toothpaste decorated with red stripes, in the early 1960s. Posters promoted it as 'the toothpaste with germ-fighting red stripes' – the crucial ingredient being an 'active concentration' of the disinfectant hexachlorophene. What was so clever about

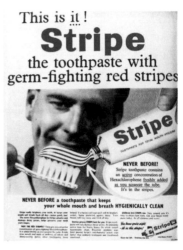

Stripe/Signal had nothing to do with hexachlorophene (manufacturers had been putting disinfectant in toothpaste since 1908), and everything to do with colour. The cheerful, vibrant red stripes were regarded as visual 'proof' of the brand's medicinal power.

Red is the most successful colour in English Premier League history. Between 1992 and 2020, a team in red won the title nineteen times. The only English clubs to win consecutive European Cups both wore red: Liverpool (1976/77, 1977/78) and Nottingham Forest (1978/79, 1979/80).

When Nottingham Forest football club was founded in 1865, they wore red-tasselled caps in honour of Giuseppe Garibaldi. For a time, Forest were nicknamed the 'Garibaldi Reds'. Visiting Britain in April 1864, the Italian freedom fighter was so popular he inspired a song with the refrain 'Hurrah for the red shirt! Garibaldi and glory.' One observer, also indelibly associated with red, was less enthused: Karl Marx described the tour as a 'miserable spectacle of imbecility'.

Garibaldi and his Reds, on the road to Italian liberation.

Garibaldi's volunteer troops, instrumental in the liberation and unification of Italy, were widely known as 'red shirts', from the distinctive feature of their ad hoc uniforms. But why did they adopt the red shirt? There are two popular theories. Some biographers trace the association back to Uruguay in the 1840s. Sentenced to death for his role in an unsuccessful uprising in Piedmont, Garibaldi fled to South America. Commanding a legion of Italian expats to defend Montevideo against a right-wing rebel army, Garibaldi and his men wore red shirts that were intended to be worn in slaughterhouses. Camouflaging the blood of men, rather than of animals, the shirts became synonymous with the Italian freedom fighter. The other explanation is that Garibaldi, who lived in New York between 1850 and 1853, was inspired by the city's volunteer firefighters, who wore red flannel shirts. It is not clear whether his troops wore red shirts when Garibaldi returned to Italy in 1848, but they certainly did in April 1860, when they landed in Sicily, to launch a campaign that would end with Italian unification.

The practice of using red to highlight crucial words in a document can be traced back at least as far as ancient Egypt, when priests used the colour to emphasise important phrases, often warning of danger. One papyrus referring to Apep (aka Apophis), a deity associated with evil, chaos and destruction, was written entirely in red. Thousands of years later, monks used the colour to give prominence to certain sections of a manuscript, often the initial capital or the opening words – this is the meaning of the word 'rubric'. The same rationale led scribes to highlight saint's days in red on the calendar, and prompted accountants to use red ink to draw attention to financial losses – from which we get the phrase 'in the red'.

Some of the lip paint worn by prostitutes in ancient Greece contained sheep sweat, human saliva and crocodile dung. The favourite lip rouge of Martha Washington, America's first First Lady, was made from wax, hog's lard, spermaceti, alkanet root, almond oil, balsam, raisins and sugar. Queen Elizabeth I's formula was slightly less noxious: cochineal, gum arabic, egg whites and fig milk.

Red lipstick is still made with cochineal, which is sourced from a tiny parasitical beetle found on prickly pear cacti in South and Central America and southern parts of north America. If you take 70,000 insects, dry them and crush them, you can extract enough red carminic acid to make an ounce of carmine dye, with the right additives. This is pretty much what the Incas, Aztecs and others did for centuries, creating a dye they used to colour clothes, houses, baskets and teeth. The Zapotec word for 'red' is the same as their word for 'colour', which indicates its importance.

Red dye from crushed cochineal bugs used for making carpets in the village of Teotitlan de Valle in the Oaxaca Valley, Mexico.

When Hernán Cortés and his cohorts conquered Central and South America and effectively enslaved the indigenous people in the early sixteenth century, they farmed cochineal so that the Spanish – who gave themselves a legal monopoly on this bright, rich and durable dye – could export it to Europe. Within decades cochineal was Spain's second most lucrative export from Mexico, after silver.

Cochineal isn't as lucrative as it used to be – the German chemical giant BASF discovered a synthetic alternative in the late nineteenth century – but it is still used in food dye, medicine (research suggests we are more likely to have faith in red painkillers) and cosmetics. The fact that cochineal beetles aren't particularly photogenic has helped prolong the trade. Imagine the outcry if pandas had to die to make lipstick.

Red, the most popular choice on national flags, can stand for the blood spilled by patriots, strength, revolutionary politics, volcanoes (on Iceland's) or a thriving carpet-making industry (the mainly red vertical stripe on Turkmenistan's flag represents a strip of carpet). The horizontal bands on Latvia's flag are of a deep carmine hue that's widely known as 'Latvian red'. This flag was replaced by a Soviet red version during the lifetime of the USSR, when some patriotic Latvians became supporters of Spartak Moscow because the football club's red–white–red strip closely resembled the original Latvian flag.

YELLOW
FEVERS

Before yellow there was gold: not a pigment as such, but a pure metallic colour, used in celebration of the divine and the royal. Images of the pharaohs were adorned with gold, a substance that in parts of the ancient Near and Middle East was regarded as having fallen to earth from the sun. The Greeks applied gold to images of their deities, and the Romans added gold to statues of their gods and heroes. The gold-skinned horses above the entrance to St Mark's cathedral in Venice used to stand in the hippodrome of Byzantium, the city in which the art of gold-leaf mosaic was first perfected. The Byzantine heritage of St Mark's is evident inside the cathedral, where the golden mosaics – like those of the ancient churches of Rome, Ravenna, Palermo and Monreale – have lost none of their lustre in the centuries since they were created.

In pre-Renaissance Italy, painters depicted Christ and the Madonna and the saints on fields of gold leaf – the incorruptible, radiant and costly metal being the most appropriate setting for

The flowing golden locks of Botticelli's *Venus*.

the holy figures. Even as the more naturalistic aesthetic of the Renaissance was coming into being, artists were still using gold in their religious paintings. Giotto, the pre-eminent artist of the Proto-Renaissance, placed his enthroned Madonna and Child against a background of gold, and gilded altarpieces were created by later artists such as Lorenzo Monaco and Fra Angelico. Botticelli added gold highlights to his *Birth of Venus*. After that, gold largely disappears from Western art, outside of the pages of monastic manuscripts, and when it re-emerges at the turn of the twentieth century, in the paintings of Gustav Klimt, it signifies sensuality and wealth rather than anything spiritual. Klimt's gold-drenched *Portrait of Adele Bloch-Bauer* was sold for $135 million in 2006, then a record price for a painting.

Klimt's notorious *Beethoven Frieze* still adorns the walls of the gallery for which it was created – the Secession Building, whose filigree golden dome is a much-photographed feature of the Vienna cityscape. The most famous golden dome in the world, however, is the Dome of the Rock, the Islamic masterpiece that illuminates the skyline of Jerusalem. The shrine was raised in the seventh century, making it one of the oldest Islamic buildings in existence, but its roof was not always the resplendent structure we see today. The Dome was rebuilt in 1022, and from then until 1964 it was a lustreless thing. As Matthew Teller notes in *Nine Quarters of Jerusalem*: 'For some 942 years it was dark grey lead. It was then renovated with sheets of anodised aluminium that looked golden, before in 1993 being re-covered in copper panels plated with a two-micron layer of actual gold.'

The further away from the equator we live – and the more rainfall we have – the more likely we are to associate yellow with happiness, according to research published in the *Journal of Educational Psychology* in 2019, based on a poll of 6,625 people in 55 countries. In Egypt, which averages 3,451 hours of sunshine a year (out of a possible 4,383), less than 6 per cent of people equated yellow with joy. In Finland, where a fleeting summer of midnight sun is more than offset by a long dark winter, almost nine out of ten people did.

When American commercial artist Harvey Ross Ball was commissioned to design a logo to cheer people up, he immediately chose yellow. Running an advertising agency in Worcester, Massachusetts, in the early 1960s, he was asked by State Mutual Life Assurance to design a symbol to boost staff morale after a recent merger.

In less than ten minutes, Ball drew a yellow smiley face, with one eye bigger than the other. He charged $45. State Mutual commissioned smiley buttons, posters and signs to encourage staff to smile while doing their jobs. Whether the scheme actually made staff happier or not, the design was deemed an instant success.

Ball didn't copyright his creation, telling his son Charles, 'Hey, I can only eat one steak at a time, drive one car at a time.' Brothers Bernard and Murray Spain, who owned two Hallmark Cards shops in Philadelphia, were harder-nosed. Adding the phrase 'Have a happy day', they copyrighted the revised design in America. By the end of 1972, they had sold more than 50 million buttons.

The joy of yellow: poet Amanda Gorman at the Biden inauguration, 2021.

In France in 1971, journalist Franklin Loufrani began using a smiley face to highlight good news in *France Soir* newspaper. In 1996, Franklin's son Nicolas made the mark an official Smiley Company brand, popularising it through global licensing agreements, some of which included the world's first graphic emoticons. The Smiley is so simple, so minimalist and effective, that it has been used to subvert, celebrate and forewarn. Dave Gibbon, the British artist who collaborated on the *Watchmen* comic book series, which made great play of a blood-drenched Smiley, said: 'It's just a yellow field with three marks on it. It couldn't be simpler. And so, to that degree, it's empty, it's ready for meaning. If you put it in a nursery setting, it fits in well. If you put it on a riot policeman's gas mask, it becomes something completely different.'

'Yellow is a colour capable of charming God.'
Vincent van Gogh

When we think of Vincent van Gogh we immediately think of yellow, and of sunflowers. His most famous images of the joyful yellow blooms were created in a building known as the Yellow House, in the Provençal town of Arles, which was van Gogh's home in 1888–89 – an especially turbulent, yet fabulously creative, period of his short life. He felt possessive about them, writing to his brother Theo: 'The sunflower is mine in a way.' Paul Gauguin, who stayed with him there for two months, captured his friend's devotion in the portrait titled *The Painter of Sunflowers*. Gauguin recalled later, 'Oh yes, he loved yellow, this good Vincent – those glimmers of sunlight rekindled his soul.'

In van Gogh's late masterpieces, colour is almost everything, used in an emotive, subjective and expressive way that prefigures

Van Gogh's *The Starry Night*, created with Indian yellow pigment.

much modern art. The most celebrated of the more than two hundred paintings he produced in Arles is *The Starry Night*, painted a little over a year before his death. Its luminous moon and swirling stars were painted using a complex pigment called Indian yellow. 'Sold in hard, dirty-coloured and ill-smelling balls', as Philip Ball notes in his book *Bright Earth*, this pigment – known as *purree, puri, piuri* or *peori* in India – was believed, by colour man George Field, to be made from animal urine (most likely cows or camels) or fluid from snakes.

In January 1883, speculation about the pigment had reached such a pitch that Sir Joseph Hooker, director of Kew Gardens, asked the India Office to clarify matters. That August, he received a report from the Indian civil servant T.N. Mukharji concluding that every ball of Indian yellow was produced by a caste of *gwalas* (milkmen) in the Bengal village of Mirzapur.

As Mukharji noted: 'They feed the cows solely with mango leaves and water, which increases the bile pigment and imparts to the urine a bright yellow colour. It is said that cows thus fed die within two years, but the piuri manufacturers assured me that this statement is wrong and, indeed, I myself saw cows six or seven years old.' The animals, he conceded, looked 'very unhealthy', probably because their owners were reluctant to vary their diet in case it made their urine less yellow. At the end of each day, the urine was heated, causing the yellow sediment to separate so it could be strained, made into balls and exported. Mukharji estimated that 'An average cow passes three quarts of urine a day, which yields about two ounces of piuri.'

However, Mukharji's account has been called into question, because there seems to be no other record of the peculiar process he describes. Whatever the veracity of his report, by 1908, according to Philip Ball, 'Indian yellow had all but disappeared.'

A naturally occurring arsenic sulphide (As_2S_3) mineral, rich yellow orpiment is present in the famous bust of Nefertiti, sculpted in 1345 BCE and also in the ninth-century Book of Kells and the walls of the Taj Mahal. 'Orpiment' is a corruption of the Latin word *auripigmentum* (*aurum* meaning 'gold') and its lustrous colour led many alchemists to waste their time trying to extract gold from it. As the mineral is 60 per cent arsenic, it probably led to as much illness as disappointment. In ancient times, many slaves died from mining the stuff, and the Italian painter Cennino Cennini warned, in his treatise *Il Libro dell'Arte* (written in the 1390s): 'Beware of soiling your mouth with it, lest you suffer personal injury.' Orpiment still plays a part in some industrial processes, but the pigment is used by very few artists.

Gamboge, sourced mainly from the solidified sap of Garcinia trees in Cambodia (the word is a corruption of the country's name) was brought to Europe by Dutch traders in 1603. The saffron-mustard pigment produced from the sap was widely used to dye the robes of Buddhist monks, and featured in the palettes of many European artists, including Rembrandt and Turner.

Gamboge can also be used as a purgative – it's so powerful that, in 1842, the Scottish physician and toxicologist Sir Robert

A contemporary satirical image of the effects of Morison's Vegetable Universal Pills.

Christison warned of 'fatal accidents which have arisen from the general employment of a notorious nostrum composed in a great measure of gamboge'. The nostrum in question was the Vegetable Universal Medicine, promoted by self-proclaimed 'hygeist' James Morison, who sold as many as ten million of his yellow Vegetable Universal Medicine pills a year. At least eleven Britons died after taking too hefty a dose and in 1836, Morison was convicted of manslaughter after one Captain Mackenzie, a 'stout, healthy man', swallowed 35 pills to ease the pain in his knee and – not to put too fine a point on it – pooed himself to death. Morison fled to Paris, where he died in 1840, a wealthy man.

The natural gamboge pigment was still in production until the early years of this century, but has now been replaced by synthesised gamboge and another artificial pigment called aureolin, which, though not quite as bright as gamboge, doesn't fade as fast.

On the evening of 6 April 1895, at London's Cadogan Hotel, Oscar Wilde was arrested on charges of committing acts of gross indecency. The writer rose unsteadily (he had been drinking heavily since hearing that a warrant had been issued), donned his coat, picked up a book with a yellow cover and was taken to Bow Street Police Station.

The book in question was *Aphrodite*, a scandalous, yellow-covered novel by Wilde's friend Pierre Louÿs. At the time, a yellow cover was a kind of code, signifying that the enclosed contents were racy, and quite likely French. (Wilde's Dorian Gray reads a deliciously 'poisonous' yellow-covered book called *Le Secret de Raoul*.) A typical newspaper headline read, 'Arrest of Oscar Wilde. Yellow book under his arm', leading to a misunderstanding which had unexpected repercussions. Many outraged citizens concluded that Wilde had been reading an issue of *The Yellow Book*, an avant-garde periodical made notorious by art editor Aubrey Beardsley's provocative and brilliant illustrations.

Wilde had never written for the journal – in 1884, he had dismissed the first issue as 'dull and loathsome' – but it was produced by his publisher, Bodley Head. After Wilde's arrest, the imprint's co-founder, John Lane, took Wilde's books off sale but an enraged mob still stoned its London office, breaking the windows

There followed an indefatigable campaign by Mary Augusta Ward – a popular novelist under the pen name Mrs Humphrey Ward – against *The Yellow Book*, persuading poet William Watson, an occasional contributor, to threaten to leave Bodley Head unless Beardsley's designs were withdrawn and he was fired. One of the art editor's greatest drawings, *The Climax*, had illustrated

The first editions of Beardsley's *Yellow Book* and *Private Eye*.

an edition of Wilde's play *Salome* and, according to art historian Kenneth Clark, had 'aroused more horror and indignation than any graphic work hitherto produced in England'.

Beardsley's dismissal became inevitable. It wasn't a decision Lane took easily, but it did mean that he no longer had to vet Beardsley's drawings for 'inappropriate details'. He tried to reposition the quarterly to appeal to a broader audience, as he had originally intended. Alas, by then the controversy had, in the eyes of many, turned *The Yellow Book* into a sleazy, decadent and homosexual brand. The last issue was published in the spring of 1897, just a few months before the celebration of Queen Victoria's Diamond Jubilee. Lane later said that Wilde's arrest 'killed *The Yellow Book* and damn near killed me'.

The first issue of satirical magazine *Private Eye*, published on 25 October 1961, was printed on mustard-yellow paper, almost certainly an allusion to 'yellow journalism', a label that was first attached to two American newspapers – William Randolph Hearst's *New York Journal* and Joseph Pulitzer's *New York World* – in the 1890s.

In one of the fiercest circulation wars ever fought, both newspapers ran a cartoon strip, drawn by Richard Outcault, that originally appeared – in black and white – in the *New York World* on 17 February 1895. One of its characters, The Yellow Kid, became so popular that Outcault gave him a bright yellow nightshirt, and the strip, expanding to a full page, became the unique selling point of the paper's colour Sunday supplement. When Hearst poached Outcault – and The Yellow Kid – for the *New York Journal*, Pulitzer commissioned another strip featuring The Yellow Kid, set in a different neighbourhood. At their peak, the rival Yellow Kids appeared several times a week and inspired the newspaper industry's first merchandising drive.

These newspapers became known as purveyors of 'Yellow Kid' journalism, a phrase which was abbreviated to 'yellow journalism' and morphed into the pejorative 'yellow press', as Hearst and Pulitzer fanned the public hysteria that drove America's invasion of Cuba, then a Spanish colony, in May 1898.

On 15 February 1898, the battleship USS *Maine* sank in Havana harbour, killing 260 American sailors. The cause of the explosion is still debated – recent research suggests it was caused by a fire in the coal bunker – but the next day the *Journal's* front-page headline read: 'CRISIS IS AT HAND, Cabinet in session, Growing belief in SPANISH TREACHERY'. Hearst did not, as urban myth has it, telegram painter Frederic Remington to say, 'You provide the pictures, I'll furnish the war', but he did offer a $50,000 reward to find the 'perpetrators' and launched a 'War with Spain' card game to boost circulation. Such sensationalism was not new – in France, one eighteenth-century scandal sheet added a footnote to one article saying 'Half of this article is true', without specifying which half – but the dishonesty of the yellow press set a dismal precedent.

In choosing its distinctive colour, the creators of *The Yellow Book* were flaunting their aesthetic refinement – a refinement taken to the point of unhealthiness. They were making a virtue of a super-sensitive sickliness, of a kind exemplified by the hero of J.–K. Huysmans' book *À Rebours* (*Against Nature*), which could be regarded as the gold standard of decadent literature. Yellow was the colour of dandyism, just as it had been during the regency of the future George IV. And nobody was ever more of a dandy than Oscar Wilde, who, on his lucrative American tour in 1882,

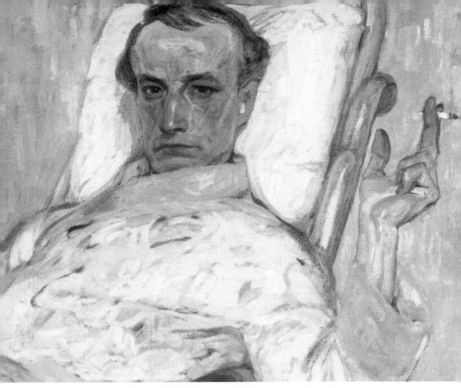

František Kupka, *The Yellow Scale* (Self-Portrait, 1907), epitomised dandyism.

brandished a yellow silk handkerchief and sunflowers so often that, as he told an interviewer, 'In Mobile [Alabama], an enterprising little boy made twenty-five dollars selling sunflowers to people who came to my lecture. That boy will be a congressman.'

In China, blue movies are often known as 'yellow videos'. In 1993, when it became legal to publish pornographic magazines in the country, they were called 'yellow books'.

Jaune, French for 'yellow', is the root of 'jaundice', an ailment whose sufferers produce so much bilirubin (the yellowish-red chief pigment of human bile) that it yellows their skin and the whites of their eyes. We probably began using 'jaundiced' to describe a bitter or cynical view of the world in the early seventeenth century. The term played off the illness, but also older, mysterious associations between yellow and ignoble emotions.

In heraldic treatises and encyclopaedias in the late Middle Ages, Michel Pastoureau reports in his book on the colour, yellow often corresponds with five particular vices (deceit, envy, hypocrisy, jealousy and treachery) and only three virtues (strength, honour and nobility). Every colour has contradictory meanings, but Pastoureau says no colour – not even black – had as many negative associations as yellow in this era.

When Charles III, Duke of Bourbon and the Auvergne, defected to the Habsburgs in 1523, it was ruled that his mansion near the Louvre should be confiscated and the doors and windows painted yellow, 'the colour of infamy'. On 24 August 1572, during the Saint Bartholomew's Day massacre, a group of Catholics, who were not content with killing, eviscerating, castrating and decapitating Huguenot leader Admiral Gaspard de Coligny, painted the doors and windows of his mansion yellow to emphasise the enormity of his crimes.

One of the great imponderables in the story of yellow is how, why and when it came to signify cowardice. In 1928, just as Hollywood was moving from silence to sound, the term 'yellow streak' was used in the epic melodrama *Noah's Ark*. Jimmy Cagney would often call his opponents out for being yellow, notably in *Taxi* (1932), when, playing a vengeful cab driver, he

David Landau, James Cagney and Loretta Young in *Taxi*, 1932.

shouts at his brother's killer: 'Come out and take it, you yellow-bellied rat, or I'll give it to you through the door.' (This is the line that is often misquoted as 'You dirty rat'). In Westerns of the period, cowards were likely to be called 'yellow-livered' or be told that their spine was yellow. Aficionados of cowboy slang have suggested that the phrase 'yellow-bellied varmint' alludes to the yellow-bellied marmot, a large species of squirrel which hides from predators in rock piles and is found all over the American West.

The *Online Etymology Dictionary* says confidently that 'yellow-bellied' was first used to signify cowardice in America in 1924. The same term was used to describe Mexican troops fighting in Texas in 1842, although opinions differ as to whether it was a racist slur, a comparison to treacherous reptiles (particularly snakes and lizards) or a commentary on their cowardice.

If you live in the north-eastern United States and want to sell your house quickly, colour consultant Sally Augustin has some advice for you: 'If you have a yellow front door, repaint it. Nobody really knows why but many realtors in the region say that it takes longer to sell a house with a yellow door.' Indeed, the prejudice against the colour seems to apply across the United States, according to online real estate marketplace Zillow. Analysis of house sales between January 2010 and May 2018 found that houses with yellow doors sell for $3,000 less than expected when such parameters as square footage, age of house, location and date of sale were factored in. (In contrast, a black front door could add around $6,000 to the price.)

'Yellow is the typically earthly colour. It can never have profound meaning … it may be paralleled, in human nature, with madness, not with melancholy or hypochondriacal mania, but rather with violent raving lunacy.'
Wassily Kandinsky, *Concerning The Spiritual In Art,* 1910

As Kandinsky's ideas about colour were somewhat idiosyncratic, it would be easy to dismiss his interpretation of yellow as further proof of his eccentricity. Yet in Russia cheap yellow paint was used to paint asylums in the reign of Catherine the Great (1762–96). The deplorable conditions made these institutions – and the slang term 'yellow house' – notorious across Tsarist Russia. The Tsarist regime also obliged sex workers to carry yellow passports.

These documents were so widely used that some Jewish women applied for them so they would be free to travel beyond the Pale, the region in western Russia where Jews were forced to settle between 1791 and 1917.

Yellow was identified with prostitutes elsewhere, too. Over the centuries, sex workers have been ordered to wear yellow scarves in Seville, Venice and Vienna, yellow headbands in Pisa, and yellow cloaks in Bergamo and Leipzig. The association was perpetuated by such artists as Édouard Touraine, who, in

Édouard Touraine's 1901 caricature of prostitutes and their clients.

a caricature published in *Le Rire* in 1901, shows young women dressed entirely in yellow.

❖ ❖ ❖

In ancient Rome, yellow was regarded as a 'female' colour. The poet Martial refers to transvestites who rent yellow dresses and recline on beds draped with yellow, and Juvenal takes a particular individual's penchant for pale yellow attire as proof of his effeminacy, hypocrisy and homosexuality. When the populist leader Clodius gatecrashed a women-only ceremony in the house of Pompeia, Caesar's first wife, he disguised himself in a saffron-dyed dress. He fled when a servant noticed his strangely deep voice.

❖ ❖ ❖

Like many songs the Beatles recorded in the late 1960s, 'Yellow Submarine' was rumoured to be about drugs. In New York, capsules of Nembutal (a brand of the barbiturate pentobarbital) were, for a time, known as yellow submarines. Paul McCartney, who wrote the ditty with John Lennon for Ringo to sing, would have none of this, saying: 'It was just a children's song. Kids get it straight away.'

In October 1966, a few weeks after 'Yellow Submarine' topped

the UK singles charts, singer-songwriter Donovan, who had contributed a couple of lines to the Beatles' number, reached the top ten with 'Mellow Yellow'. Donovan's famous composition is variously said to have been inspired by getting high on fried banana skins (a claim

since proven to be a hoax), vibrators (known as electrical bananas in the 1960s, and mentioned in the lyric) and James Joyce's description of Molly's buttocks in *Ulysses* ('He kisses the plump mellow yellow smellow melons of her rump').

Interviewed years later, Donovan said the song meant: 'Quite a few things. Being mellow, laid back, chilled out. "They call me mellow yellow, I'm the guy who can calm you down."' You didn't have to take drugs to reach the desired level of mellowness, he insisted, you could just meditate. Only the year before, in another hit song 'Colours', Donovan had celebrated mellow as the feeling he got when he saw his yellow-haired love in the light of early morning.

If you want to avoid being bitten by a shark, do not wear 'safety yellow'. Although it is often worn by coastguards, the hue is supposedly so attractive to sharks that some divers call it 'yum-yum yellow'. In 2008, tests by the Discovery Channel, using bait in different coloured bags, found that the yellow ones were attacked by sharks slightly more often than the other colours. Other researchers have concluded that all high-contrast colours – yellow, orange and red – are likely to attract the attention of certain species of sharks, simply because they are more visible.

Local authorities in the town of Holland, Michigan, proudly claim that the Yellow Brick Road, which leads Dorothy and her pals to the Emerald City in Frank L. Baum's novel *The Wonderful*

Dorothy and Toto head out of Kansas on the Yellow Brick Road.

Wizard of Oz (1900), was inspired by a yellow-bricked street that ran near Baum's summer cottage in their neighbourhood. This explanation is disputed by residents in Bradford, Pennsylvania, where Baum ran a print shop in the 1870s – they insist he was inspired by the creamy pale yellow bricks produced by local businessman William Hanley. Both origin stories are a tad prosaic. When you are creating a magical kingdom of winged monkeys, Munchkins, emerald cities and ruby slippers, why shouldn't the road be yellow?

Just as many Europeans can claim to be descended from Charlemagne, many Chinese believe they have a common ancestor, Xuanyuan Huangdi, better known as the Yellow Emperor, who, if he existed at all, reigned around 2697–2597 BCE.

When Taoism emerged in China, in the fourth century BCE, at its heart was the concept of Wuxing (Five Phases), which sees all phenomena in terms of the interaction of five elements: fire, wood, water, metal and earth. The reign of the quasi-legendary Xuanyuan Huangdi was understood to have been governed by earth, the element signified by the colour yellow.

Jing Han, an expert on dyes and social status in imperial China, says the first emperor officially recorded to have worn yellow was Wen of Sui, who reigned from 581 to 604 CE. Under the succeeding dynasty, the Tang (618–907), the imperial monopoly of yellow was justified with the argument that, as there could never be two suns in the same sky, there could only be one emperor. The rules, Han notes, became more elaborate under the Qing dynasty (1644–1912): 'Bright yellow could only be used for the court robes and dragon robes of the emperor and empress. The robes for crown princes should be apricot yellow and those for other princes should be golden yellow.'

With minor royals ordered to wear blue, this scheme ensured that the emperor and empress would stand out at their own court. The royal prerogative was gradually undermined by the emperors themselves, who began granting bodyguards the right to wear bright yellow. The Dowager Empress, who controlled the Chinese government from 1861 until her death in 1908, caused a scandal by giving her favourite train driver a yellow jacket.

The toppling of the Qing dynasty in February 1912 effectively ended imperial yellow's symbolic power. The new republic's national flag had five coloured stripes representing the major ethnic groups: red to signify the Han majority, black for the Hui, yellow for the Manchus, blue for the Mongols and white for Tibetans. In 1949, Mao's Communist government replaced this with a Soviet-style flag, with one large yellow star and four smaller yellow stars on a red background. The large star represents the Communist Party's guiding role and the smaller ones, which initially stood for the four revolutionary classes – the proletariat, the peasants, the petty bourgeois and 'patriotic capitalists' – were later said to symbolise the country's main ethnic groups.

'Imagine that awful being and you have a mental picture of Dr Fu-Manchu, the yellow peril incarnate in one man.'
Sax Rohmer, *The Insidious Dr Fu-Manchu*, 1913

Sax Rohmer, the creator of Dr Fu-Manchu, was hardly the first novelist to invoke the menace of the 'yellow peril'. A squalid opium den featured in Charles Dickens' unfinished novel *The Mystery of Edwin Drood* (1870), and the Chinese community in London's East End was one of the principal targets of British Sinophobia. The American scare was more far-reaching. In 1873, in response to the influx of immigrant cheap labour, the *San Francisco Chronicle* warned: 'The Chinese invasion! They are coming, 900,000 strong.' Such sentiments paved the way for the Chinese Exclusion Act (1882), which banned emigration to America from China and was not completely dismantled until 1943.

In his autobiography *Timebends*, playwright Arthur Miller wrote: 'The Hearst press was periodically frantic about an oncoming "Yellow Peril", with the Tong Wars in Chinatown as proof that the Chinese were bloodthirsty, sneaky and ... lustful for white women. Many were the front pages with the immense black headlines "Tong Wars" – accompanied by drawings of Chinese cutting each other's heads off and holding them up victoriously by their pigtails.'

The clan violence of the Tong Wars raged in various Chinatowns across the USA, particularly in San Francisco, from 1880 to 1913. By the time they fizzled out, the world was on the brink of a conflict caused, to a significant degree, by Kaiser Wilhelm II, who had publicly urged European powers to confront the 'yellow peril'. The Kaiser did not coin the term – Russian sociologist Jacques Novikov used it at the close of the nineteenth century to describe the West's fear of China – but his concern was sparked by Japan's humiliating triumph over Russia in 1905, the first time in modern history that an Asian nation had defeated a major European power in war.

Brazil's national football team used to wear white shirts with blue collars. After failing to win the 1950 World Cup in their own backyard, they blamed their kit, which – the Rio newspaper *Correia da Manhã* declared – suffered from a 'psychological and moral lack of symbolism'. In response, the team adopted a strip that mirrored the blue, yellow and green of the Brazilian flag. That standard had been designed in 1817 by Jean-Baptiste Debret during the reign

Brazil's first World Cup triumph, resplendent in Habsburg yellow, Sweden, 1958.

of Dom Pedro I, the first monarch to rule an independent Brazil, who was married to Maria Leopoldina, daughter of the Habsburg Emperor Francis II. The new flag dutifully combined yellow – representing the Habsburg dynasty – and the green associated with the Braganza dynasty. More than two centuries later, Brazil's football team still wear Habsburg colour shirts.

In 1981, when the American hostages returned from Iran, they were welcomed by a profusion of yellow ribbons. That greeting was inspired by the song 'Tie a Yellow Ribbon Round the Ole Oak Tree', a global bestseller for Tony Orlando and Dawn in 1975 in which a returning convict hopes the love of his life will put up these ribbons to symbolise her forgiveness.

That song was partly inspired by the legend that the wives and fiancées of soldiers and officers in the US Cavalry wore a yellow ribbon as a symbol of constancy and devotion, especially during the Civil War. Yet researchers at the American Folklife Center found no evidence of such a practice in contemporary records. American folklorist Gerald E. Parsons concluded that yellow's association with loyalty was largely a product of the success of the great John Ford/John Wayne western *She Wore A Yellow Ribbon* (1949) and its popular theme tune.

Cycling is seldom far from controversy, so it seems appropriate that the origin of the sport's most important icon, the *Maillot Jaune* ('Yellow Jersey') worn by the race leader in the Tour de France, is still disputed.

Here is the conventional account. On 19 July 1919, before the eleventh stage of the Tour de France set off from Grenoble, founder and organiser Henri Desgrange urged race leader French cyclist Eugène Christophe to wear a yellow jersey to make it easier for competitors, journalists and spectators to recognise him. Desgrange's real motive was probably to promote his sports paper, *L'Auto*, which was sponsoring the race and was printed on yellow newsprint. A cover illustration for *Vie au Grand Air* magazine, dated 19 July 1919, shows Christophe in a mustard-yellow jersey. Spectators allegedly laughed at him 'because he looked like a canary'.

The *Maillot Jaune* makes its first appearance in print on the 1919 Tour.

Yet in 1957, talking to *Champions et Vedettes* magazine, Philippe Thys, the Belgian cyclist who actually won the 1919 Tour, claimed that he had reluctantly worn the yellow jersey back in 1913. In his account, his team manager, Alphonse Baugé, pressured him to wear the conspicuous garment in order to please the team's sponsor, Peugeot. Jacques Augendre, the Tour's official historian and archivist, and for many years its organiser, noted that Thys was a 'valorous rider ... well known for his intelligence', whose memories 'seem free from all suspicion', before concluding: 'No newspaper mentions a yellow jersey before the war [and] being at a loss for witnesses, we can't solve this enigma.'

The simple explanation for the global popularity of yellow taxis – as seen on the streets of Chicago, Derby, Hartlepool, Kolkata, Melbourne, New York and Rio de Janeiro, among others – is that they are more visible than other colours. In 1907, Chicago car salesman John D. Hertz launched his Yellow Cab Company, followed in 1912 by Albert Rockwell's Yellow Taxicab Company in New York, but neither Hertz nor Rockwell was the first to brand taxis in this colour. In Paris, in the mid-1820s, a yellow horse-drawn carriage may have given Eugène Delacroix the inspiration he needed to proceed with his *The Execution of the Doge Marino Faliero*.

His friend Alexandre Dumas describes how Delacroix called a cab to take him to the Louvre, where he intended to study some pictures by Rubens, in order to decide how best to use yellow in his painting's complex colour scheme. 'Delacroix stopped short before the body of his cab: it was yellow like that which he wanted, but in the position where the carriage was placed, what gave it that dazzling tone? It was not the tone itself, it was the

In praise of taxis: Eugène Delacroix's *The Execution of the Doge Marino Faliero* (1826–27).

shading which made it come out. But these shadings were violet. Delacroix had no further occasion to go to the Louvre; he paid the cab and went upstairs to his room: he had caught his effect.'

The visual effect that had so struck the artist was an example of 'simultaneous contrast', in which our perception of a colour is modified by an adjacent colour – a phenomenon explored at length by the chemist and colour theorist Michel Eugène Chevreul a few years later.

'Each Jew, after he shall be seven years old, shall wear a badge on his outer garment that is to say in the form of two tables joined of yellow felt of the length of six inches and the breadth of three inches.'
Statute of the Jewry, 1275, passed in the reign of King Edward I

A short, brutish, man in a yellow cloak stares angrily into the face of the friend he is about to betray. This confrontation is central to Giotto's fresco *Kiss of Judas*. The work, part of a cycle portraying Jesus's life and death, was commissioned in the first decade of the fourteenth century by the banker Enrico Scrovegni for the chapel adjoining his palace in Padua. As moneylending – usury – was still officially a sin in the eyes of the Church, the chapel and its frescoes may well have been a grand gesture of expiation.

In Giotto's masterpiece, Jesus's arrest in Gethsemane is presented as the act of a lynch mob. Judas looks like a thug in a hurry. He also wears a yellow cloak in another fresco in Giotto's series, collecting his thirty pieces of silver. Although the rogue disciple was not consistently colour coded in medieval art, Judas had often been depicted in yellow since the twelfth century, and it's probable that Giotto's use of that colour was reinforcing the idea that the Jews were responsible for the murder of Jesus – Judas wears yellow, and by the time these frescoes were painted yellow had long been used to identify, isolate and persecute Jews.

In the eighth century CE, during the reign of Caliph Umar II, the Islamic empire decreed that every dhimmi (non-Muslim)

Giotto's fresco *Kiss of Judas* (1304–06), Scrovegni Chapel, Padua, Italy.

should wear a particular colour to identify which minority they belonged to. Jews were obliged to wear honey-mustard. It is impossible to say definitively why yellow was chosen, but if you're going to make people wear something to identify their ethnicity, it makes sense for that colour to be as visible as possible.

In France in 1269, an edict from King Louis IX instructed Jews of both sexes to wear a yellow cloth in the shape of a wheel, concluding: 'If a Jew is found without this insignia, the upper part of his clothing will belong to the one who encountered and discovered him thus.' His son, Philip III, used the decree to fill the royal coffers, commanding tax collectors to sell the badges at a fixed annual price. In England, the Statute of Jewry, passed under Edward I in 1275, likewise stipulated that any Jew over the age of seven should wear a yellow badge. In 1278–79, 278 Jews were hanged in England, resentment against them having been stirred by the very activity – moneylending – that financed Giotto's frescoes.

'The authorities have warned that severe punishment – up to and including death by shooting – is in store for Jews who do not wear the yellow badge on their back and their front.'
The *Judenrat* (Jewish Council) in Bialystok, June 1941

Adolf Hitler became Chancellor of Germany on 1 January 1933. Three months later, during a boycott of Jewish businesses, SA and SS members painted the Star of David in yellow and black, or yellow and red, on thousands of shop windows across Germany. The yellow badge was to become one of the emblems of the Nazi genocide.

By the end of 1939, Jews in Nazi-occupied Poland had been variously ordered to wear a white band, a blue and white star of David, a yellow band, a yellow badge and a yellow Star of David. On 14 November 1939, SS brigade leader Friedrich Uebelhoer directed that all Jews in the Kalisz district in central Poland should wear a 10-centimetre-wide band, beneath their armpit, in a shade he identified as *judengelber Farbe* – literally 'Jewish-yellow colour'.

There was never a single 'Jewish yellow-colour' but the marking was applied across the Nazi empire. On 1 September 1941, Reinhard Heydrich, director of the SS's Reich Main Security Office, decreed that all Jews living in the Reich, aged six years or older, must wear a yellow Star of David on a black field. Croatia's pro-Nazi Ustashe regime adopted the yellow badge in May 1941, applying it with genocidal zeal. The Bulgarian government, bowing to German pressure in August 1942, introduced a small yellow button, which most Jews never wore. Exporting the yellow badge to Western Europe proved more problematic. The Danes ignored it, the Dutch protested and many shocked French citizens regarded the star as a relic of the Middle Ages. Even though 72,500 Jews were deported from France to death camps, German agents in Bordeaux, Nancy and Paris reported that people were wearing yellow handkerchiefs and yellow scarves – and carrying yellow flowers – to show solidarity with the Jews.

INTO THE BLUE

Why do we love blue so much? In a 2015 YouGov study, blue was the most popular colour in all ten countries surveyed, favoured by 23 per cent of respondents in Indonesia, 26 per cent in China and more than three out of ten Americans, Britons and Germans. Although the factors that influence our colour preferences are still the subject of intense debate, that finding has been echoed in other studies.

How do we explain blue's popularity? There are many theories as to why we like or dislike certain colours. Some scientists argue that colours once acted as useful signals helping us to adapt our behaviour as we were evolving as a species (by, for example, distinguishing ripe fruit from rotten fruit), and that these associations still influence us. Others suggest that we simply like the colours that make us feel good – which means that we tend to prefer bright and light colours to darker and heavier ones. In 2010, Stephen E. Palmer and Karen B. Schloss made a compelling case for the ecological valence theory, which suggests that we like the colours we associate with our favourite environments, objects and experiences. As colour consultant Sally Augustin puts it: 'To our ancestors, blue in nature signified a good day, whether it was the blue sky on a sunny day or the blue of a calm sea. We subconsciously associate blue with calmness, coolness, tranquillity and a sense of security – we feel we can trust a blue sky and blue sea.'

This might help to explain why forty-five of the world's hundred largest banks have made blue integral to their brand. Even in China, where red is a nationalistic, politically correct and lucky colour, seven of the largest banks use blue.

A phenomenon known as scattering is the reason a clear sky looks blue. When sunlight strikes the Earth's atmosphere, it causes the electrons and protons of the atmospheric gases (mainly nitrogen and oxygen) to oscillate, which in turn causes a scattering of the light. The blue component of the visible spectrum, having shorter wavelengths and higher frequencies than the red, creates faster oscillations, which means that the blue is scattered more efficiently. Hence we see a predominantly blue sky above us. Violet-indigo light is scattered even more strongly than blue, but there's less violet than blue in sunlight, and our eyes are much less sensitive to it.

A blue sky is paler towards the horizon because oblique light has to pass through a much thicker belt of the atmosphere than light that is coming from overhead. Accordingly, the photons of every part of the spectrum get scattered many times over, which has the effect of tilting the balance away from blue. With the dominance of blue light diminished, we see an increase in the amount of white light. At sunset, when the light source is at its

Scattering in opalescent glass: it appears blue from the side, but orange light shines through.

lowest point in the sky, the longer-wavelength red and yellow light prevails over the blue, which is scattered to such an extent as to have become, in effect, diluted.

Why is the sea in Homer's *Odyssey* and *Iliad* described as *oínopa pónton* – a phrase meaning something like 'wine-faced', but most famously translated as 'wine-dark'? Why is it not blue? Why is nothing in Homer ever blue? These questions were first addressed in detail by classical scholar and future prime minister William Gladstone, in his book *Studies on Homer and the Homeric Age* (1858). Observing that Homer frequently used the adjectives black and white but rarely made use of red, yellow or green, and that the total avoidance of blue was also a feature of other ancient Greek texts, Gladstone surmised that 'the organ of colour and its impressions were only partially developed among

A wine-dark sea: Odysseus enduring the Sirens – third-century CE mosaic.

the Greeks of the heroic age'. Goethe, in his *Theory of Colours*, had similarly concluded that there was something defective in the Greeks' perception of colour. Their literature seemed to be almost monochromatic.

Philologist Lazarus Geiger proposed that the Greeks were not alone in this respect. After studying a plethora of ancient writings, including the Icelandic sagas, the Koran, and texts in Chinese and Hebrew, he pronounced that blue was absent from all of them. German ophthalmologist and medical historian Hugo Magnus had a straightforward Darwinian explanation for the omission: the human race had evolved since the age of Homer. Whereas our distant ancestors could distinguish only between red, orange and yellow, the modern eye is able to discern blue and violet.

The idea is nonsensical, of course. Why would the ancient Greeks have valued lapis lazuli so highly, for example, if they couldn't see its colour? The reason that the language of Homer has no direct equivalent of 'blue' is a matter of linguistics, not physiology. It is simply a different way of describing the world. Gladstone was on the right track when he argued that the ancient Greeks used light/dark contrasts when categorising colour, rather than hues. Where we might apply the label 'blue' to the ocean, they might use the epithet *glaukos* (the origin of glaucoma), a colour-neutral term that originally meant 'gleaming' or 'shining' and came to mean 'grey'. Or they might use *kyaneos* (for darker non-vivid shades), or *porphyreos* (generally signifying vivid shades ranging from blue to violet to ruby), or *lampros* (for a metallic-silvery-azure), or *melas* (for the darkest tones).

The world is perpetually in motion, rather than offering us a panorama of fixed colours, and the ancient Greek language reflects that mobility. The sea at dawn can be silver one day and pink the next. At sunset it could be blood-red or black. And sometimes it might have a dark depth that puts you in mind of a rich red wine.

Why do we 'feel blue', or say that 'we have the blues'? The modern phrase first appears in English in Francis Grose's 1785 *Classical Dictionary of the Vulgar Tongue*, which defines 'to look blue' as 'to look confounded, terrified or disappointed', and by the mid-nineteenth century there are a number of references to 'feeling blue' and 'to have the blue devils' among American writers. The association of blue with melancholy is probably connected to our lips turning blue after death, though one theory is that it has nautical origins, as sailing ships that lost their captain would fly blue flags and paint a blue band along their hull when returning to their home port. But it's those 'blue devils' that seem more resonant. They are associated with the visual hallucinations of alcohol withdrawal, and are commonly thought to be the root of the word for blues music. Some US states, even now, have 'blue laws' prohibiting alcohol sales on Sundays.

Colonial America's so-called 'blue laws' reinforced strict religious standards, prohibited frivolity on the Sabbath and regulated (or prohibited) the sale of alcohol. They later spread across America and, as they primarily concerned the Sabbath, became known as 'Sunday laws'. The explanation of the term may be alochol withdrawal, as above, though historian Patrick J. Mahoney attributes the term to eighteenth-century religious dissenters who deplored the rigidity of New England Puritanism. He quotes the scholar James Hammond Trumbull's observation that, in the rhetoric of the period, 'to be "blue" was to be "puritanic", precise in the observance of legal and religious obligations, rigid, gloomy, over-strict'.

Traces of woad, a dye extracted from the leaves of the wildflower *Isatis tinctoria*, have been found in a Neolithic cavern in the Bouches-du-Rhône department in the south of France, in Iron Age sites in Britain and Denmark, and in Egypt, where it was first used as a dye in around 2500 BCE. It's more closely associated with the Picts, who inhabited northern and eastern Scotland from the first century BCE to the tenth century CE, and are reputed to have painted themselves blue with woad. As historian Tim Clarkson writes: 'The word "Picti" is a Latin term meaning "painted people". Its Roman origin suggests that it was a nickname given by troops on the imperial frontier to barbarians lurking in the wild lands beyond.' In *The Conquest of Gaul*, Julius Caesar informs his readers that the whole island is inhabited by

'Blue Gold' – crushed woad leaves in a workshop at Lectoure, France, reviving the traditional woad industry of the region.

body-daubers: 'All the Britons paint themselves with woad which produces a dark blue colour and for this reason they are much more frightful in appearance during battle.' Historians believe that Queen Boudicca and her Iceni warriors smeared themselves with it in their rebellion against the Romans, which began in 60 BCE. But, despite what Mel Gibson would have us believe, the Scottish rebel William 'Braveheart' Wallace (1270–1305) and his supporters did not paint their faces with woad. Or wear kilts.

By the Middle Ages, woad had become one of the most common textile dyes in Europe, alongside weld (yellow) and madder (red). Making woad, as Philip Ball writes, was not a pleasant process: 'Basically, you mashed up the plant in water and let the mixture ferment. To help it ferment, it was typically mixed with urine and left out in the sun. Sometimes the mixture would be trampled underfoot to help release the dye. And not surprisingly, it stank.'

The smell was one of the reasons Elizabeth I decreed in 1585 that no one could sow woad within four miles of a market or clothing town, and within eight miles of one of her properties. The popularity of the dye contributed to a famine in 1586, as farmers stopped growing cereals to capitalise on demand. One year after the famine, restrictions were imposed on the acreage that parishes and individuals could give over to the sowing of woad.

With competition from imported indigo and the discovery of various synthetic blues, woad gradually fell out of favour. In recent years, though, it has had something of a small-scale revival, thanks to the realisation that woad is a natural antibiotic, and can be produced without polluting the planet. Woad has re-emerged as a dye for textiles and an art pigment, and its oil is being used in organic cosmetics. East Anglia, Boudicca's old stamping ground, has long been associated with woad, so it seems fitting that the only British farm producing the dye commercially is near Dereham in Norfolk.

The expression 'true blue' – meaning to be loyal and steadfast – may have originated in Coventry's textile industry. The phrase 'as true as Coventry blue' has been traced back to 1377, and refers to the woad dye produced locally, which was renowned throughout Europe for its durability. No samples of this blue cloth have been found, however, and neither have any recipes survived, so we don't know which secret ingredient made the dye so steadfast.

On Christmas Eve 1968, William Anders, Frank Borman and James Lovell, the crew of Apollo 8, became the first people to

'Earthrise' – the 'Blue Marble' of the Earth – captured by the Apollo 8 mission.

orbit the moon. During one orbit, Anders took one of the most influential photographs in history: the Earth, half in shadow, above the lunar horizon. That picture, dubbed by its author 'Earthrise' and catalogued by NASA more prosaically as AS8-14-2383HR, revealed the Earth to us as a blue orb, partially shrouded in white cloud – a planet of majestic but fragile beauty. Four years later, the crew of Apollo 17 took the first photograph of the whole Earth, known to the world as the 'Blue Marble' and to NASA as AS17-148-22727.

The nickname of the north-west Indian city of Jodhpur – the Blue City – is inspired by the old city where, as David Abram reported in *Wanderlust* magazine, the buildings form 'an extraordinarily compact, convoluted jumble of cubes, painted every conceivable shade of blue, from deep-sea indigo to sun-bleached

Jodphur – the 'Blue City' of Rajasthan, India.

cobalt'.

According to local legend, the colour designated houses owned by Brahmin (upper-class) families, but it's likelier, as Abram notes, that 'the blue dye, derived from copper sulphate, was added to the conventional white limewash to prevent infestation by termites, and that the practice became fashionable over time'.

'Speak unto the people of Israel and bid them that they make them fringes in the borders of their garments and throughout their generations, and that they put upon the fringe of the borders a ribbon of blue.'

Book of Numbers 15:38, King James Version

In 1638, at Greyfriars Kirk in Edinburgh, thousands of aristocrats, church ministers and ordinary Scots signed a covenant to defend the Presbyterian Church of Scotland against reforms that had been proposed by King Charles I, who was insisting that an English-style prayer book be used in Scottish churches. Many Scots, convinced that no man – not even a king – could lead a church that was ruled by God, signed this National Covenant and, taking their inspiration directly from the Book of Numbers, showed their allegiance by wearing blue caps and marching behind blue banners. Their choice of colour was not entirely divine – varieties of blue had formed the backdrop of the Scottish flag for centuries and blue woollen caps were already popular in the Highlands.

Ironically, blue caps with a white cockade would later become a symbol of the Catholic Jacobite rebellions. The association is preserved in nostalgic songs about the revolts, notably 'Blue Bonnets Over the Border', possibly written by Sir Walter Scott, who in later life was not averse to wearing a blue bonnet himself.

Perhaps the first aristocratic blue bloods were the ninth-century Castilian nobles, who raised their arms to show off their blue veins – thereby emphasising their racial purity – before going into battle against the Moors. Later, all over Europe, pale skin and blue veins drew a visible distinction between the aristocracy and the masses, whose skins were tanned because they toiled outdoors.

In ancient Egypt, the only stable blue pigment was made from lapis lazuli, a rare stone first mined in Afghanistan in around 4,000 BCE. It was extremely expensive, and therefore the preserve of the upper classes. Made by mixing the pulverised powder with animal fat or vegetable gum, the thick blue paste was used to honour the regal dead. Lapis lazuli was found in the inlaid eyebrows and make-up on Tutankhamun's mask, for example. Living royals wore it, too – Cleopatra's eye shadow was based on a fine lapis lazuli dust – and it was widely used for other decorative purposes. Around 2,600 BCE, the first known synthetic pigment, the bright Egyptian blue, was created by heating a mixture of silica, copper, lime and an alkali such as potash.

Face of the ancient Egyptian goddess Hathor, lapis lazuli, c.1100 BCE.

In the Middle Ages, making many pigments was an arduous business, but the mixing of ultramarine (literally 'beyond the sea') was particularly complicated – as Cennino Cennini (1360–1427) underlines in his craftsman's handbook *Il Libro dell'Arte*. You needed top-quality lapis lazuli stones (which were rare and expensive because they were almost entirely mined in Sar-e-Sang in Badakhshan in northern Afghanistan), and a lot of skill, judgement, muscle power and patience: 'Then take a pound of the powder of lapis lazuli; mix it all well together into a paste, and that you may be able to handle the paste, take linseed oil, and keep your hands always well anointed with this oil. This paste must be kept at least three days and three nights, kneading it a little every day.' This is just one stage of the process. Luckily, in Cennini's view, it was worth it: 'Ultramarine blue is a colour illustrious, beautiful and most perfect, beyond all other colours ... Let some of that colour, combined with gold, which adorns all the works of our profession, whether on wall or on panel, shine forth in every object.'

The costly ultramarine was the perfect colour with which to create images of the Virgin Mary, the Queen of Heaven, whose status as *Theotokos* ('God-bearer') was confirmed at the First Council of Ephesus in 431 CE. As Katy Kelleher observed in the *Paris Review* in 2016, 'After Mary's status as a deity incubator became canon, artists started creating even more portraits of the holy mother and under the court of Constantinople, these pieces began to take on a rather standardised look – she was shown dressed in Marian blue-cloth against a flat backdrop of gold leaf, holding the infant Jesus with a serene expression on her face.'

For the Virgin Mary's attire, ultramarine was the pigment of choice. If you could not afford it, you could mix lapis lazuli with other cheaper pigments or use azurite, a deep blue mineral produced by the weathering of copper ore deposits.

In medieval Europe, the cult of Marian blue was so powerful that French king Louis VII (1120–80) tried to use the colour to save the dynasty. His first marriage to the wealthy and powerful Eleanor of Aquitaine failed to produce a male heir and was annulled after the Second Crusade. Louis then adopted Marian blue – and the golden fleur-de-lis, a symbol associated with Charlemagne – for cloth, banners and furnishings, thereby setting the template for the French monarchy's coat of arms. In 1165, his third wife – his second having died in childbirth – gave him a son, who, as Philip II, made France the richest, most powerful country in Europe.

Even after the revolutionary rupture of 1789, and the final collapse of the monarchy in 1870, blue remained one of France's defining colours. It features (alongside red and white) on the national flag and the naval ensign, occupies half of the standard of the capital, Paris (alongside red), and is worn by the national rugby and football teams.

Most of France's former colonies have expunged blue from their flags, although Quebec's Fleurdelisé (Lily-Flowered) still consists of a white cross on a blue background with four white fleurs-de-lis.

'What is blue? Blue is the invisible becoming visible. Blue has no dimensions. It is beyond the dimensions of which other colours partake.'
Yves Klein

A synthetic ultramarine was developed in the 1820s by two chemists – Jean-Baptiste Guimet in France and Christian Gmelin in Germany. Although artists as diverse as Paul Cézanne, Vincent van Gogh, Georges Braque, Jackson Pollock, Georges Seurat, Amedeo Modigliani and Piet Mondrian all took to the pigment, synthetic ultramarine disappointed Yves Klein, who found it dull and dark, lacking the luminosity of its organic predecessor. He spent much of his short life – he died of a heart attack in 1962, at the age of 34 – striving to recapture the magic of ultramarine.

Klein created his first 'artwork' – actually a gesture made in conversation with friends – in 1947, when he was 19, 'signing' the blue sky above his native Nice. In 1955, the Salon des Réalités Nouvelles rejected an orange monochrome picture that Klein had submitted. One member told Klein's mother: 'If Yves would

Visitors contemplate an untitled work by Yves Klein at the Guggenheim Bilbao's 2005 retrospective.

accept to add at least a little line, or a dot, or even simply a spot of another colour, we could show it. But a single colour, no, no, really that's impossible.'

Klein would prove the Salon gloriously wrong. Collaborating with colour merchants and chemists to develop a synthetic binder, Rhodopas M60A, he eventually created a pigment that he called International Klein Blue (IKB), and used almost exclusively, as a painter and sculptor, for the last five years of his life. Pure and intense, IKB has a velvety matte surface and looks, as Victoria Finlay put it, 'deep enough to swim in'. Gaze at a painting like *IKB 79* for a significant length of time and you may understand why artist Michael Craig-Martin said: 'The power of a single blue painting to stay in one's imagination for one's lifetime – that's quite something.'

In his documentary *Blue* (1993), Derek Jarman tells us: 'Blue is the universal love in which man bathes, it is the terrestrial paradise.' His film consists of a static shade of International Klein Blue accompanied by a soundtrack of music and narration. As

Jarman contemplates the AIDS epidemic, the 'fathomless colour of bliss' also becomes the colour of death: 'The virus rages fierce. I have no friends who are not dead or dying. Like a blue frost it caught them.'

Great artists love a challenge – and when Thomas Gainsborough read that his rival Joshua Reynolds had said that blue was too cold to be used as the predominant colour of a painting, he set out to prove him wrong. The result is one of the most famous of all British paintings, *The Blue Boy* (1770), in which he used ultramarine, cobalt, slate, turquoise, charcoal and indigo to make the boy's satin coat look iridescent.

In his book *Concerning the Spiritual in Art* (1910), Wassily Kandinsky paid eloquent tribute to the power of blue: 'The inclination of blue towards depth is so great that it becomes more intense the darker the tone, and has a more characteristic inner effect. The deeper the blue becomes, the more strongly it calls man towards the infinite, awakening in him a desire for the pure (and finally) the supernatural ... Blue is the typically heavenly colour. Blue unfolds in its lowest depths the element of tranquillity.' He preferred the colour's darker tones – he likened them to the notes produced by the cello or double bass that lead us into a 'profound state of seriousness'.

If one event could be said to have been the source of Picasso's so-called Blue Period, it was the suicide in Paris of his friend, the 20-year-old painter Carles Casagemas. Depression, addiction and impotence had tormented Casagemas for some time. After announcing his imminent return to Barcelona, he invited some friends to a farewell dinner on 17 February 1901 at the Café de l'Hippodrome. There he shot himself.

Although the two artists had set up a studio as students in Barcelona, been virtually inseparable for eighteen months, and discussed launching an art magazine together, they were estranged when Casagemas died. A month before, on a trip to Málaga, Picasso had become so fed up with his troubled friend, he sent Casagemas back to Barcelona. It has been suggested that, before returning to Paris, Casagemas unsuccessfully sought a reconciliation. If that's so, Picasso's Blue Period might have been underpinned by guilt as much as grief.

The picture generally considered to mark the beginning of this period was *Casagemas in his Coffin*, which was completed later in 1901. For the next three years, most of Picasso's work depicted prostitutes, beggars and drunks, in a palette consisting almost entirely of blue and bluish greens. In one of the last works of the Blue Period, *La Vie* (1903), Casagemas is again the subject, naked except for a loincloth, as a naked woman clutches him.

Prussian blue, the pigment that dominates *La Vie*, was created accidentally by German chemists Johann Jacob Diesbach and

Johann Konrad Dippel in 1704, when a batch of cochineal red turned blue after a chemical reaction with some animal oil in the potash they were using. The first modern synthetic pigment, Prussian blue was much easier to make than ultramarine, and far less expensive, at around one tenth of the price. Commercially available from around 1724, it was initially favoured by artists at the Prussian court (hence the name), but its intense hue soon made it popular with a host of foreign artists, notably Antoine Watteau and François Boucher. Hokusai's iconic woodblock, *The Great Wave off Kanagawa*, is a fine demonstration of the power of Prussian blue, even though the colour has faded over the years.

Unlike some synthetic pigments created in the eighteenth century – such as the notorious Scheele's green – Prussian blue was not toxic. Indeed, it is used in Germany, Japan and the United States in very small doses to treat radiation poisoning. Patients who have been treated with Prussian blue for a long time have reported that it turned their tears and sweat blue.

Other synthetic blues followed the creation of Prussian blue: cerulean blue in 1805, cobalt blue in 1807 and artificial ultramarine in 1824. There was then a bit of a wait until the 1920s, when the vivid – and immediately popular – phthalocyanine blue (or phthalo blue) was invented.

In March 2021, YInMn Blue, a pigment with all the vibrancy of ultramarine, became commercially available. Rather like Prussian blue, YInMn came, as it were, out of the blue. In 2009, Mas Subramanian, a professor of material science at Oregon State University, was trying – with graduate student Andrew

YInMn Blue – the world's newest pigment, synthesised by Mas Subramanian.

E. Smith – to create an inorganic material for use in electronic devices. When one sample came out of the furnace brilliantly and intensely blue, Subramanian (who had previously worked for DuPont) immediately recognised the substance's potential as a pigment. YInMn – the name comes from the chemical components' symbols on the periodic table: yttrium, indium and manganese – can be made without toxins (unlike, say, cobalt), is very stable and so good at reflecting infrared that it could be used to cool exterior surfaces. Unfortunately, because of the rarity of yttrium and indium, YInMn Blue is extremely expensive.

In 1847, when Mauritius became the first British colony to issue stamps, it printed its twopenny (2d) stamps with ink containing

Prussian blue and ultramarine. Only twelve of these stamps are known to exist. Being worth an estimated £1 million at auction, they are are known as 'the Crown Jewels of philately'.

❖ ❖ ❖

Marc Chagall's stained-glass windows in All Saints' Church in the Kent village of Tudeley are a poignant reminder of blue's power to move us spiritually, emotionally and aesthetically. Local landowners Sir Henry and Lady d'Avigdor-Goldsmid commissioned the first window in memory of their 21-year-old daughter Sarah, who had drowned at sea in 1963. Two years earlier, Sarah and her mother had marvelled at the windows that Chagall had created for the Hadassah Medical Center in Jerusalem, when they were on display in the Louvre.

Working with Reims stained-glass master Charles Marq, Chagall developed a special technique for applying pigment directly to the glass, which gave his work its particular radiance. The first window depicts the events of Sarah's death, but also shows her resurrected soul rushing towards Christ's embrace. Attending the window's unveiling in Tudeley's church in 1967, Chagall said: 'It's magnificent, I will do them all.' The twelfth – and last – window was

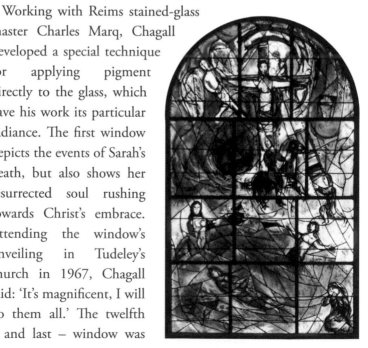

fitted in 1985, the year of his death. In all of them, a gorgeous cobalt blue predominates.

Blue can be heavenly, but it can also be profane, as in blue movies, blue language (swear words) and blue humour. There are many explanations for blue's seedier aura. None are conclusive. In 1859, slang expert John Camden Hotten suggested that the association originated with popular books printed on cheap blue paper in France in the seventeenth and eighteenth centuries. Yet the *Oxford English Dictionary*'s view is that 'such material appears in general to have been highly moral in tone'.

We do know that in 1824, in the *Scottish Gallovidian Encyclopedia*, John McTaggart defined 'thread of blue' as 'any little smutty touch in songwriting, chatting or piece of writing', without giving any clue as to where this usage originated. It has been variously suggested that the connection with sex might have something to do with convicted prostitutes being obliged to wear blue in sixteenth-century England or, rather fancifully, with the blue walls of Chinese brothels. The association may have been reinforced by the use of blue spotlights in burlesque shows in the late nineteenth century.

In England, blue humour was popularised by stand-up comedian Max Miller (1894–1963), who liked to ask his music hall audiences which book they wanted him to read jokes from: his 'clean' white book or his 'dirty' blue one. As the self-styled Cheeky Chappie, Miller made a career out of his blue book. A typical Miller joke would begin: 'When roses are red, they're ripe for plucking, when a girl is sixteen she's ready for …', at which point he would pause, let the audience fill in the blank and shout 'Ere!' in mock outrage.

'I remember the dinner party he [Alfred Hitchcock] gave one night for about 12 people. When you entered the house, you noticed that there were blue flowers on the front hall table. As you went in, everything was blue. Blue covers on the tables, the steaks were blue, the knives and forks were blue.' That was how, in a 1989 interview, actor James Stewart recalled one of the oddest dinners he ever had in Hollywood. Stewart didn't know that Hitchcock had already held a blue dinner party – at the Trocadero in London in honour of the actor Sir Gerald du Maurier and Broadway star Gertrude Lawrence. On that occasion, even the brown rolls were blue when broken open.

The *Oxford Dictionary of Word Origins* states that 'blue devils' were 'originally baleful demons punishing sinners. In the eighteenth century, people fancifully imagined them to be behind depression'. This usage was still current in the late nineteenth century, but guitarist Debra Devi, author of *The Language of the Blues*, suggests that the musical genre's name is rooted in a different kind of 'blue devils' – it was a term for the hallucinations that beset alcoholics during an attack of delirium tremens.

On plantations in the Mississippi Delta, black slaves sang work songs that were, as folklorist Alan Lomax wrote, 'the powerful, bitter poetry of a hard-pressed people' – music we would now categorise as blues. In 1903, while waiting for a train in Tutwiler, Mississippi, the African-American bandleader W.C. Handy came across just such a hard-pressed man. In his autobiography, *Father of the Blues*,

Handy recalled: 'A lean, loose-jointed Negro had commenced plucking a guitar beside me while I slept. His face had on it some of the sadness of the ages. As he played, he pressed a knife on the strings of a guitar in a manner popularised by Hawaiian guitarists who used steel bars. The effect was unforgettable.' It was, wrote Handy, 'the weirdest music I ever heard'.

The refrain refers to the junction of two railway lines in Muirhead, 42 miles south of Tutwiler, Mississippi. Inspired and intrigued, Handy began writing and performing such songs as 'Saint Louis Blues'. His breakthrough instrumental, 'Memphis Blues', published in 1912, was so popular in dance halls across America that 'blues' became a marketing term. Published as a southern rag, 'Memphis Blues' sounds more like big band jazz than raw blues. Handy's memorable foot-tapper mixes conventional 16-bar melodies with the 12-bar sections that

W.C. Handy's classic 'Saint Louis Blues', a staple for Louis Armstrong.

became synonymous with the blues and contains, Handy wrote, 'blue notes – the transitional flat thirds and sevenths … by which I was attempting to suggest the typical slurs of the Negro voice'.

The Illustrated London News records the 1829 race between Cambridge in light blue and Oxford in dark.

In 1829, William Gladstone watched the very first boat race between Oxford and Cambridge. Oxford's rowers wore a dark blue because that was the colour of Christchurch, the college at which five of the crew were studying. This is the shade we now call Oxford blue. Cambridge competed in white, with most of its crew wearing the pink ties of St John's College as sashes. Seven years later, just before the second race, Cambridge were about to wear white until, as Walter Bradford Woodgate notes in his history of the contest, a certain Mr Phillips dashed off for a 'a bit of Eton ribbon (light blue) for luck.' It worked. Cambridge won by 20 lengths.

But Eton's signature blue – a pale shelduck hue (now officially designated as Pantone 7464C) – is not the same as Cambridge blue (Pantone 557C), which is classified as 'spring green' on the RGB colour wheel. The greening of Cambridge blue is often attributed to Alf Twinn, university boatman between 1934 and 1984, who, either because he wanted to distinguish the rowing colours from the rugby union team's, or because his eyesight was deteriorating, began adding subtle amounts of yellow to the blue paint used for oar blades.

Why do Italian national sports teams mostly wear blue, rather than one of the colours of the national flag? The answer is that the blue shirts honour the House of Savoy, under which the country was unified in 1861. Italy's football team are known globally as *gli Azzurri*, but they wore white shirts in their first match, against France, in May 1910. Unable to agree on a colour, they chose to compromise on the no-colour of white. By January 1911, when they lost 1–0 to Hungary in Milan, the authorities had decided that blue was the colour.

Written by Richard Rodgers and Lorenz Hart in 1934, 'Blue Moon' has been covered countless times, but perhaps the bluest rendition is the one by Elvis Presley, recorded at Sun Records in 1956. Simon Schama thinks so. In an essay on the colour blue for the *Financial Times* in 2017, he wrote: 'If you want the all-time moody howl, there's only one performance to have in your head while you're hunting for blue suede shoes. That would be "Blue Moon", sung, to the soft clip-clop of a cowboy's horse, by Elvis: the only version of the song which drops Lorenz Hart's unpersuasively golden ending.'

We normally have one full moon a month, but, as a lunar cycle lasts only 29.5 days, we sometimes get a thirteenth moon, which we call a blue moon. The Royal Observatory suggests the term may be a mispronunciation of the disused word 'belewe', which meant 'to betray' – reflecting the fact that, centuries ago, the extra moon would have made it harder to determine when (and for how long) Lent would last. If you ever see a moon that's actually blue, it will be because of dust particles in the atmosphere, usually a product of a volcanic eruption.

The tablets on which Moses received the Ten Commandments on Mount Ararat were of blue stone, according to the Jewish Talmud, a tradition illustrated by Marc Chagall in his painting *Moses Receiving the Tablets of Law* (1956).

The Hindu gods Krishna, Rama, Shiva and Vishnu are usually depicted with blue skin to symbolise calmness, courage, intuition and transcendence. In the ancient religious text called the *Brahma*

Krishna dancing and playing flute – Chitra Mahal, Bundi, Rajasthan, India.

Samhita, Krishna is described as having a body the colour of a bluish cloud. In other texts, he is likened to a lotus flower with blue and white petals.

If the bluebird did not sing, according to Iroquois mythology, we would be living in perpetual winter. The species' birdsong is credited with driving off Tawiscaron, the destructive demigod of winter. The Pueblo and Navajo tribes associate the bluebird with the sun. To the Cherokee, it represents the wind and has the power to predict, if not control, the weather. One Native American folk tale says that the bluebird's feathers, originally an ugly colour, became blue after it swam four times in a lake, which no rivers flowed into, singing at the top of its voice.

The bluebird of happiness is a trope in Chinese mythology, Russian fairy tales and the folklore of Lorraine, the last of these being the obvious inspiration for the Belgian poet and playwright Maurice Maeterlinck's 1908 play *The Blue Bird*, in which brother and sister Tyltyl and Mytyl seek happiness, symbolised by the bird.

The play's popularity is credited with promoting the 'bluebird of happiness' myth in the English-speaking world. It has been suggested that lyricist Yip Harburg was alluding to Maeterlinck when he wrote the line 'Somewhere over the rainbow, blue birds fly' for Judy Garland to sing in *The Wizard of Oz*. The MGM musical prompted Twentieth Century Fox to adapt *The Blue Bird* into a movie starring Shirley Temple. Shot in black and white and Technicolor, the musical fantasy took some liberties with the play, making Mytyl (Temple) a precocious brat who needed taking down a peg or two.

Maeterlinck knew Oscar Wilde and may have been inspired by the latter's 1891 essay, 'The Decay of Lying', a Socratic dialogue in which Vivian dreams of a world where 'over our heads will float the Blue Bird, singing of beautiful and impossible things, of things that are lovely and that never happen, of things that are not and that should be'.

Long before it became synonymous with denim, indigo, a pigment originally extracted from the plant *Indigofera tinctoria* and related species, was widely used as a dye in Egypt, India, Japan and Peru. From about 300 CE, the Maya mixed indigo with the clay mineral palygorskite and heated the blend to create the bright, distinctive and almost indestructible pigment known as Maya blue, with which they coloured frescoes, pottery and people sacrificed to the rain god Chaak.

Maya blue frescoes of musicians from the site of Bonampak, Chiapas.

The Mayan civilisation collapsed in the eighth and ninth centuries but indigo's deadly history lasted another thousand years. When fermenting, the indigo leaves produced toxic fumes that attracted swarms of flies and killed many planters (effectively slave labourers) in Latin America and the Indian subcontinent. In 1858, in Bengal, a campaign of passive resistance against the East India Company's ruthless management of the crop led a British government commission to conclude that 'not a chest of indigo reached England without being stained with human blood'. Mercifully, by the end of the nineteenth century, companies such as BASF had begun manufacturing synthetic indigo.

'Navy blue' is such a ubiquitous term that it seems odd even to consider its origins. But they are British and quite specific. The dark blue colour of naval uniforms was adopted for officers of the British Royal Navy in 1748 and gradually spread to navies worldwide. The colour (originally termed 'marine blue') comes from the indigo plant and its particular seafaring appeal is that it is colour fast when exposed to sun and salt water. Indigo also dyes silk, cotton and wool with equal efficiency. Its arrival in the mid-eighteenth century was no accident, for it coincided with the expansion of the British East India Company's control of the trade and subsequent colonisation. A little ironically, the British Navy modified its uniform requirements for those actually serving in the tropics, who from 1877 were designated to wear all-white tunics and trousers ('bush dress').

Without indigo, there would be no denim. When Nevada tailor Jacob Davis designed a pair of sturdy working men's trousers in

1870, he wrote to his fabric supplier Levi Strauss: 'The secret of them Pents is the Rivits that I put in those Pockots'. Davis and his supplier patented the design in 1873, but by the time they were being sold as Levis it was clear that the secret of their success was indigo dye, not rivets.

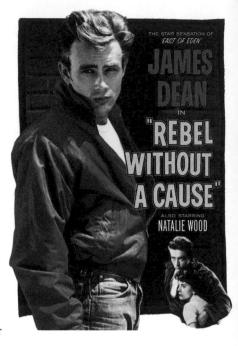

Unlike most dyes, indigo did not penetrate the cotton yarn – it sat on top of it. With wear, tear and time, the molecules of dye fell away, causing each pair of jeans to fade in a completely different way, as if life itself was acting as a bespoke tailor.

Jeans have, of course, become one of America's most enduring and successful cultural exports, right up alongside the Big Mac, thanks to the accumulative and successive efforts of, among others, cowboys, Hollywood (in Westerns and, later, as emblems of rebellion in the films of James Dean and Marlon Brando) and GIs stationed abroad.

The phrase 'thin blue line' – often used to describe the police's role as the last defence against crime – became prominent in 1988 with Errol Morris's documentary *The Thin Blue Line*, about the murder of a Texas police officer, but the colour has been synonymous with law and order for centuries. In 1829, the first

Metropolitan Police officers in London wore blue swallowtail coats to distinguish themselves from the red uniforms worn by the military police. Although some say the nickname bluebottle is derived from Cockney rhyming slang – 'bottle and glass' means 'arse', which makes the police 'blue arses' – in Shakespeare's *Henry IV Part II*, the prostitute Doll Tearsheet denounces the beadle (a kind of minor functionary responsible for making sure people appear in court) as a 'blue-bottle rogue'.

The blue flower became one of the central motifs of German Romanticism with the publication in 1802 of *Heinrich von Ofterdingen*, a posthumous and unfinished novel by Georg Philipp Friedrich Freiherr von Hardenberg, aka Novalis. His blue flower is a complex and fluid symbol, representing a yearning for the unattainable and the infinite. In Novalis's case, it's also the emblem of his fiancée Sophie von Kühn, who died of tuberculosis when she was just 15. Novalis's own death, at the age of just 28, is usually said to have been caused by tuberculosis as well, but medical opinion now tends to blame cystic fibrosis. His life is explored brilliantly in Penelope Fitzgerald's novel *The Blue Flower* (1995).

Sometimes a flower is more than just a flower – as Norbert Hofer well knew when he wore the blue cornflower during his ultimately unsuccessful campaign for the Austrian presidency in 2016. Many Austrian politicians wear flowers in their buttonholes and, officially, the Freedom Party (and Hofer) chose the blue cornflower because it looked good. And yet, as Vienna historian Bernhard Weidinger told the BBC: 'The cornflower is a complicated symbol. It was Kaiser Wilhelm II's favourite flower and was used by pan-German nationalists in the nineteenth century. Then, between 1934 and 1938, when the Nazis were banned in Austria, it was the secret symbol they wore to recognise each other.' In 2017, about to enjoy a brief spell in government as the junior partner in a coalition, the Freedom Party tried to soften its image by replacing the cornflower with the edelweiss, Austria's national flower.

A blue rose cannot be produced naturally, because roses can't synthesis delphinid, the pigment that makes most blue flowers

Suntory's genetically-modified blue rose, 'Applause'.

blue, but plant breeders at the Japanese company Suntory have tried to create a blue rose through genetic modification. The project has not progressed much since 2010, when a variety called Applause went on sale, at a princely £24 per stem. It arguably looks more mauve than blue.

The tiny roundworm *Caenorhabditis elegans* has no eyes, but it does have an aversion to blue. They wiggle away from it even though they have only the most basic system for sensing light. When they are foraging in compost, they need to avoid the brilliant, lethal, blue toxin produced by *Pseudomonas aeruginosa* bacteria. A study by American physiologist Dipon Ghosh, published in *Science* magazine in March 2021, found that the worms were much slower to turn away from the toxin in darkness than in light. Presented with a mutated beige strain of the bacterium, the worms reacted the same in light and dark and were generally slower to wiggle away than when the toxin was blue. Ghosh concluded: 'The worms aren't sensing the world in a grayscale and evaluating levels of brightness and darkness. They're actually comparing ratios of wavelengths and using that information to make decisions.' In other words, they are using other parts of their cellular system to help them see.

HOUSE OF ORANGE

On 11 July 1976, Barbara Marx became the fourth and final Mrs Frank Sinatra. Soon after the wedding, she began, as David McClintick wrote in *Architectural Digest* magazine, 'a gentle war on orange'. Sinatra's belief that 'orange is the happiest colour' was reflected in the décor of the modest house in Rancho Mirage, Palm Springs he owned from the mid-1950s until 1995. As McClintick observed: 'There was orange carpeting and orange tile, an orange refrigerator and orange draperies, orange towels, and an orange sofa.' In two photographs taken at this home – with his labrador Ringo and friend Yul Brynner – Sinatra is wearing an orange shirt and brown trousers, a colour combination he returned to later when he took up abstract art as a hobby.

Barbara Sinatra and her friend Bernice Korshak, a Beverly Hills interior designer, replaced much of the orange in the living room, bar and projection room with subtler desert colours and whites. The singer approved the makeover after his wife shrewdly suggested that every room should be named after one of his songs.

Frank Sinatra's Rancho Mirage home at Palm Springs.

Even Korshak's 'gentle war' could not completely eradicate the colour from Sinatra's Palm Springs compound. The caboose (named 'Chicago'), latterly the main hangout for the singer, friends and family, was still exuberantly orange when he sold the place. In his refurbished bedroom – 'I Sing the Songs' – the double bed retained its orange and white fabric headboard.

In the 1980s, Sinatra took up painting seriously, producing Abstract Expressionist works inspired by Mark Rothko, Robert Mangold and Ellsworth Kelly. He was so proud of his 1984 orange, brown and white abstract painting *Desert* – the original of which hung in the projection room in Palm Springs – he had prints made for friends.

The word 'orange' is derived from *naranga*, the Sanskrit word for orange trees, which are plentiful in India. (Najpur, in central India, is still known as Orange City.) In other words, the term described a fruit long before it defined a colour. The accepted theory is that from *naranga* we get *narang* in Persian, *naranj* in Arabic and, possibly because it was misheard, *orenge* in old French and *orange* in English.

The colour term came into everyday use because of the fruit, which was introduced to Spain by the conquering Moors in the eighth century but not imported to Britain on any scale until the fifteenth century. In Old English, the colour was called *geoluhread* ('yellow-red'). In the 1390s, the fox that appears in a rooster's nightmare in Geoffrey Chaucer's 'Nun's Priest's Tale' is described as having a 'colour betwixe yelow and reed'. In Shakespeare, orange is not an independent colour – it's always conjoined with 'tawny'. In *A Midsummer Night's Dream*, Bottom sings of a blackbird with an 'orange tawny bill'. Whenever he uses orange as a stand-alone word, it refers to the fruit.

The popularity of the fruit – initially some Europeans were so confused they called them 'golden apples' – encouraged the more widespread use of 'orange' as a colour term in English in the seventeenth century. It finally achieved official recognition as one of Sir Isaac Newton's seven colours of the rainbow.

❖ ❖ ❖

What did Sinatra see in orange? Painter, synaesthete and colour theorist Wassily Kandinsky argued, in his book *Concerning the Spiritual in Art* (1910), that 'Orange is like a man, convinced of his own powers – red brought nearer to humanity by yellow.'

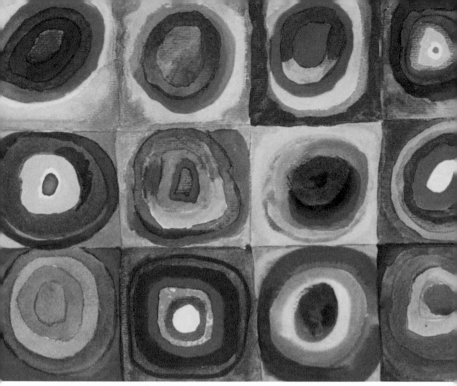

Wassily Kandinsky, *Color Study, Squares with Concentric Circles* (1913).

American agency ImagiBrand says the colour has these attributes: 'Adventurous, risk-taking, vibrant, flamboyant, stimulating to the senses, affordable, warm, sociable, optimistic, enthusiastic, cheerful, self-confident, independent, extroverted and uninhibited, creative flair, warm-hearted, agreeable and informal.' Sinatra would have been less impressed by the agency's assertion that orange could also be 'superficial, insincere, dependent, over-bearing, self-indulgent, exhibitionist, pessimistic, cheap, unsociable, and overly proud'.

In 'Why I Am Not a Painter', Frank O'Hara admits, 'One day I am thinking of a color: orange. I write a line about orange', but ends up with a poem called 'Oranges' that doesn't mention the colour at all. O'Hara's charming poem, inspired by a visit to abstract painter Mike Goldberg, stands up rather better than Christina Rossetti's 'Colour', which after reminding us what is pink, red, blue, white, yellow, green, concludes, disappointingly: 'What is orange? Why, an orange, Just an orange.' As British poet Claudia Daventry says, 'She probably wrote this, the most obvious poem about colour, because there was no rhyme for orange. She mentions colours all the way through and pulls up an image for each but the poem is strangely dead – I think because all the images are clichés and there is no work for our imagination to do.'

In 1809, Johann Wolfgang von Goethe excluded orange from his colour wheel. The parts of Goethe's wheel that look orange are called red-yellow, which he liked because its 'warmth and gladness' represented 'the intenser glow of fire', and yellow-red, which he loathed because it was popular among 'savage nations', liable to enrage animals and, worst of all: 'I have known men of education to whom its effect was intolerable if they chanced to see a person dressed in a scarlet cloak on a grey day.'

The prolific American colour theorist Faber Birren was ambivalent about orange. On the one hand, those who shared Sinatra's taste in colour were, he said, 'good-natured, likeable and social ... with an easy smile and a talent for small talk'. On the other hand: 'Chances are better than even that you will not marry – and that if you do your marriage will be one of light affection. You seldom indulge in serious thought or severe discipline.' Birren did recommend orange to corporate and government clients, primarily as a hazard warning or a therapeutic tonic, but, as a latter-day puritan with a small 'p', he preferred subtler shades in art, possibly influenced by his father, landscape painter Joseph Birren, whose colours hovered uneasily between tasteful and insipid.

The saffron shade of orange is integral to three eastern religions. In Buddhism, it symbolises the highest state of perfection – and, in its lighter shade, is the colour of the robes worn by monks who belong to Theravada, the religion's oldest existing order. In Hinduism, it is the fire that burns away ignorance and is worn by ascetics, saints and those seeking the light, as a symbol of their Guru Nanak comes upon the Sanyasi Dattetreya - Unknown, Kashmir School - Google Cultural Institute abstinence from

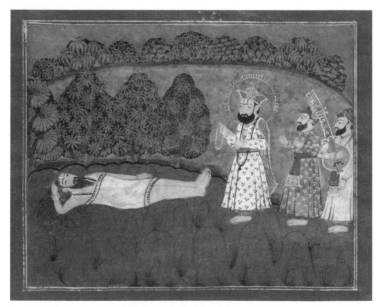

Guru Nanak, the founder of Sikhism, comes upon the Sanyasi Dattetreya. The painting is from the Kashmiri school, c.1800.

material things and pleasures. The colour has become synonymous with Hindu nationalism, and acts of violence by nationalist extremists has prompted political opponents, particularly in the Congress Party, to coin the term 'saffron terror'. Long regarded a colour of joy, community and connection by the Sikhs, a thoroughly orange shade of saffron has become a brand of protest for farmers in the Punjab protesting against the Indian government's drive to deregulate agriculture. At their rallies, the farmers and their supporters, who are mainly Sikh, wave saffron banners and their flag, the Nishan Sahib, which shows the Sikh religion's insignia in blue on an orange saffron background.

In the West, orange has become synonymous with the 'orange glow' complexion sported by overenthusiastic self-tanners – known as tanorexics – and those who slap on a lotion to simulate a sun-kissed glow. This look is especially popular with reality TV stars and Donald Trump. In August 2017, in Osnabruck in north-west Germany, an Austrian father and son were arrested after a traffic stop found five bags filled to the brim with thousands of carrot-coloured ecstasy pills made in the likeness of Donald Trump. It was reported that the drugs were being promoted on the dark net with the slogan 'Trump makes partying great again'.

Agent Orange is the most notorious of a rainbow of defoliants the American military dropped on Vietnam in the 1960s and 1970s. Others were named green, pink, purple, blue and white.

There were three kinds of Orange, but they were clearly deemed insufficiently deadly, so another more powerful variant, variously known as Enhanced Orange, Orange Plus or Super Orange, was developed by Dow Chemicals. It has been estimated that as many as three million Vietnamese people have suffered illnesses and disabilities caused by Agent Orange contamination.

As a brand, orange has also signified classic Penguin Books, record label RCA, the supermarket Sainsbury's, budget airline EasyJet, Flymo lawnmowers (which were blue and white until 1977, when it was decided that orange was easier for gardeners to see), the US Signal Corps, fashion house Hermès (forced to adopt orange during the Nazi occupation of France because it couldn't find boxes of any other colour), Princeton University (since 1928) and the telecom giant formerly known as Microtel. In 1994, Microtel's brand consultants Wolff Olins, inspired by the colour's purposeful associations in feng shui, created an orange square to dovetail with the renamed company's slogan 'The future's bright, the future's Orange'.

In 1874, Claude Monet's painting *Impression, Sunrise* made its debut in a show known retrospectively as the Exhibition of the Impressionists. The name of the painting and its style – in the eyes of critic Théodore Duret, Monet was not depicting actual landscapes as such, but the 'fleeting appearances which the accidents of atmosphere present to him' – ushered in an artistic revolution. Orange was absolutely central to Monet's ground-breaking painting – the colour of the sun, much of the sky, and the reflective surface of the sea.

Monet's *Impression, Sunrise*, 1874.

Impression, Sunrise did not impress everyone – painter, print-maker and playwright Louis Leroy sneered that 'Wallpaper in its embryonic state is more finished than this seascape' – but Impressionism became an unstoppable force, driven by rebellious artistic genius, the increased availability of bright synthetic pigments and the invention of paint tubes by American portrait painter John G. Rand in 1841. Pierre-Auguste Renoir said: 'Without colours in tubes, there would be no Cézanne, no Monet, no Pissarro and no Impressionism.'

If you wanted to escape the studio's claustrophobic confines and paint *en plein air*, as many Impressionists did, the sunny tones of orange and yellow came into their own. Chrome orange (introduced in 1809) is crucial to Renoir's *La Yole* (*The Skiff*), in which he used five other synthetic pigments – chrome yellow, cobalt blue, lead white, lemon yellow, viridian. Monet made use of another synthetic pigment, cadmium orange. The natural pigment orange ochre – popular back in Roman times – was used subtly by Camille Pissarro in *Côte des Boeufs at L'Hermitage* (1877) and Edgar Degas in *After the Bath, Woman Drying Herself* (1890–95).

The pigment realgar has such a delightfully evocative linguistic source – it comes from the Arabic *rahj-al-gâr*, meaning 'dust of the cave' – it seems a shame that, as an arsenic sulphide, it was extremely toxic. Yet before the invention of synthetic pigments many artists felt it was the only way they could create a bright orange. Like orpiment, which it closely resembles, realgar is rarely used today but in sixteenth-century Venice it was applied by those masters of colour Bellini, Tintoretto, Titian and Veronese. To achieve the subtly beautiful orange of Joseph's coat in *The Holy Family with a Shepherd* (1510), Titian employed realgar on its own for the brightest highlights and mixed it with earth pigments for the darker areas.

The Dutch fondness for orange begins with one of Europe's oldest aristocratic dynasties. Back in the eighth century, William of Gellone (known in Provence's Occitan language as Guilhem d'Aurenga) was given the Provence town we now call Orange by Charlemagne, his cousin and ally, thereby founding the house of Orange. William repelled an army of 100,000 Moors at Narbonne in 793 before retiring to the monastery he had helped found. Such piety, bravery and chivalry earned him canonisation and a starring role in the *chansons de geste*, the epic poems sung by French minstrels in the Middle Ages. One of the very first *chansons de geste* was all about him: the *Chanson de Guillaume*.

When the dynasty ran out of heirs in 1544, the title of Prince of Orange passed to an 11-year-old cousin, William, whose family owned lands in Nassau (now part of Germany). To claim his inheritance, the boy had to become a Catholic – at that time Protestants in Provence were being persecuted at the behest of France's King Francis I. As a ward at the court of Charles V, Holy Roman Emperor and King of Spain, William became so trusted that, in 1559, Charles's son Philip II made William *stadtholder* (effectively governor) of the Spanish Netherlands. As William of Orange – his taciturn ways also earned him the nickname William the Silent – he was the first leader of the revolt that led to Dutch independence, which was eventually secured in 1648.

The Prince's Flag, a horizontal tricolour of orange, white and blue, is believed to have become associated with William in the 1570s. It became the official flag of the Dutch Navy in 1587 – three years after William's assassination – and inspired the flag of New York City and the first flag of South Africa. But the orange dye faded quickly and is said to have been hard to spot at sea, a serious drawback for a maritime mercantile power. By the 1630s, the orange band was yielding to red. The red, white and blue tricolour became the Dutch national flag in 1937.

The most famous Dutch Oranje is the shade worn by the national football team since 1907. (The Pantone reference is 16-1462 TCX.)

Dutch duo: Ruud Gullit (left) celebrates after scoring with teammate Marco van Basten (12) during the UEFA Euro 88 Final.

The colour's iconic zenith, in sporting terms, came in 1974, when the Dutch team, led by the genius of Johan Cruyff, reached the World Cup final, only to lose 2–1 to West Germany. Known as Clockwork Orange, a nickname taken from Anthony Burgess's 1971 novel in acknowledgement of the precision of their play, this side thrilled purists with a style called Total Football that seemed certain to reinvent the game. It didn't quite turn out that way, but the orange worn in that defeat remains one of the most resonant football jerseys of all time, more prestigious even than the geometric-patterned shirt worn by the Dutch team – graced by such greats as Marco van Basten, Ruud Gullit and Frank Rijkaard – that won the European Championships in 1988.

As a sporting colour, orange remains a minority taste. In baseball, it is one of the colours – along with white and black – worn by the San Francisco Giants. In British football, it is worn by an eclectic assortment of clubs (Barnet, Blackpool, Dundee United,

Hull City, Luton Town and Newport County). In 2002–03, Glasgow Rangers wore an orange away strip but dropped it 'for commercial reasons' at the end of the season. In a city still fiercely divided by religion – Rangers has traditionally been regarded as a Protestant club and their rivals Celtic as Catholic – choosing a colour associated with the fervently Protestant Orange Order was criticised as 'needlessly provocative' by a Rangers shareholder. Eighteen years later, the club returned to orange, but only as lettering on a black away kit.

Orange acquired sectarian significance in 1688 when the Protestant William of Orange, William the Silent's great-grandson, invaded England, deposed his father-in-law, the Catholic King James II, and installed himself as William III, initially ruling alongside his wife, Mary. William's rout of James's army at the Battle of the Boyne on 1 July 1690 – the last time two British monarchs confronted each other on a battlefield – left behind a fragile peace. In the 1780s, as an incentive to Ireland's Catholic majority to remain loyal to the British crown – and not be distracted by America's successful war of independence – ministers eased some of the restrictions that had been imposed on them. In reaction, the Orange Order was founded in 1795.

Since then, the deeply conservative Orange Order has had a pervasive and controversial influence on the political situation in Northern Ireland and beyond, and has been criticised for its perceived closeness to Protestant paramilitary groups. In recent years, leaders have recognised that the organisation has an image problem, with the Rev Mervyn Gibson admitting: 'There are those who condemn the Order, they say it's too working class, it's archaic, it's Neanderthal and it's got a knuckle-dragging image.'

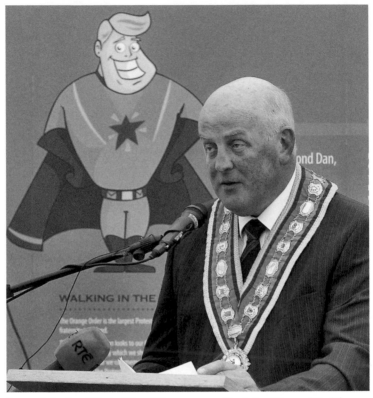

Orange Order Grand Master Edward Stevenson, speaking below 'Diamond Dan' the Orange Man, at the opening of a Museum of Orange Heritage.

In 2008, the Order tried to rebrand itself, partly by launching the superhero Diamond Dan the Orange Man (the name honours the fact that the organisation had been founded in 1795 at the Diamond farmhouse in Loughgall, owned by supporter Dan Winter). This makeover was not a runaway success, but you can still buy Diamond Dan branded fridge magnets and jigsaws on the Order's website.

An organisation that seriously intended to jettison its 'knuckle-dragging image' might have been better off rethinking the marches it stages every 12 July to celebrate William III's victory. These events have defined public perceptions of the Order so strongly across the world that an entire Marvel cinematic universe of orange superheroes couldn't change them.

Apart from the fruit, orange has become synonymous with another staple food, carrots. The romantic notion that Dutch farmers turned their tubers orange in honour of their royal family is, alas, a myth. It's true that carrots were usually purple or yellow when they were originally cultivated in Afghanistan, but in the sixth century the *Juliana Anicia Codex* (a Byzantine copy of a text by the Greek botanist Dioscorides) features a tuber – labelled 'Staphylinos Keras' ('cultivated carrot') – with a straight, orange root that looks very much like the carrots we eat today. Just over four centuries later, a cookbook for the Islamic Caliphate highlighted three cultivated varieties of carrot (*jazar*): *ahmar* (red-orange), yellow and white. In the words of the World Carrot Museum: 'Orange carrots came first, Dutch nationalism second.'

The Afrikaners, almost three million of whom still live in South Africa and Namibia, are largely descended from Dutch settlers, and the home country's colours have long had emblematic significance for them. Employees of the Dutch East India Company, which operated under a flag that closely resembled the Dutch national flag, were responsible for naming the Orange River, the longest river in the Republic of South Africa. The river gave its name to – and defined one of the borders of – the Orange Free State, which was an independent Boer republic between 1854 and

1902, before it became a province of the British colony of South Africa. During the apartheid era, the name must have sounded fiercely ironic to the black majority. 'Orange' was dropped from the name when the African National Congress took power, and it is now simply called Free State.

In 1928, keen to settle an acrimonious debate over a new national flag, the South African government adopted the Prince's Flag (the horizontal tricolour of orange, white and blue) with three flags – the Union Flag and the standards of Orange Free State and the former South African republic – displayed in miniature in the white space. As the Orange Free State's flag itself featured a miniature version of the Dutch flag, it became the only national flag to contain a flag within a flag within a flag. The Prince's Flag was chosen because, according to local legend, it was the first standard raised on South African soil by Jan van Riebeeck, the navigator who commanded the Dutch Cape colony from 1652 to 1662.

The national flag did not change when South Africa declared independence in 1961. However, after negotiations between the African National Congress and Ppresident F.W. de Klerk's government, it was discarded as a symbol of the country's colonial and racist past in 1994.

The first flag that might be described as Ireland's national flag was probably green with a golden harp on it. Known as the Green Flag, it was flown by Irish leader Owen Roe O'Neill in

1642. In the 1790s, as the standard of the United Irishmen movement (led by a Protestant, Wolfe Tone), the green mainly signified rebellion. By the 1870s, the colour had evolved into the brand of Irish nationalism. Oscar Wilde's mother Jane teased her elder sister Emily Thomazine Warren, a fervent Unionist married to a British Army officer, by writing to her on green notepaper.

The green, white and orange tricolour, adopted by the Irish Free State in 1921, was first flown in 1848 by Thomas Francis Meagher, leader of the radical Young Irelanders movement, who had been given it by French revolutionaries, in celebration of their overthrow of the monarchy. In Meagher's fine words: 'The white in the centre signifies a lasting truce between the orange and the green and I trust that, beneath its folds, the hands of the Irish Protestant and the Irish Catholic may be clasped in generous and heroic brotherhood.'

During the Easter Rising, the ill-fated revolt against the British in 1916, Irish rebels flew at least three flags: the Green Flag (emblazoned with the legend 'Irish Republic'), the green, white and orange tricolour, and the Starry Plough, a golden plough, with black and silver stars, set against the dark green backdrop of a field. The tricolour became the Irish Republic's flag during the Irish War of Independence (1919–21), and was officially recognised as such in 1937. Yet many songs that celebrate

This was the flag hoisted by the rebels over the General Post Office in Dublin in the 1916 Easter Rising.

Ireland's revolution – such as 'The Dying Rebel' – indulge in poetic licence, praising the 'green, white and gold' flag.

For followers of Indian self-styled mystic Bhagwan Shree Rajneesh in the 1970s and early 1980s, life was intensely orange. Most sannyasins – as the movement's initiated 'seekers of the truth' were known – wore orange (but also purple and red), because Rajneesh instructed them to wear the sacred colours of sunrise and sunset. In India, orange is also strongly associated with Hindu ascetics.

Rajneeshis in the midst of 'dynamic meditation' at the cult's ashram in Oregon, where the events of the Netflix series *Wild, Wild Country* took place.

In his book *My Life in Orange*, English journalist Tim Guest, whose mother Anne became a sannyasin, characterised his unconventional childhood as 'somewhere in between *Peter Pan* and *Lord of the Flies*'. He recalled: 'Sannyasins were known as "the orange people" ... everyone who came to sit around on beanbags in our living room wore only orange, wore orange dungarees, orange drawstring trousers, orange sandals, orange clothes.'

At its peak, the organisation ran 500 sannyasi centres in 32 countries. Rajneesh moved to America, selling himself as a 'Beat Zen' and effectively taking over the village of Antelope, Oregon, with 2,000 followers. The movement began to unravel in the mid-1980s amid allegations of fraud, sexual abuse, attempted murder and poisonings, a process chronicled in the 2018 Netflix documentary series *Wild, Wild Country*. In 1984, Ma Anand Sheela, Rajneesh's personal assistant, organised the biggest biological attack in US history, when – in a bid to fix local elections – the 'orange people' poisoned 751 locals by contaminating salad bars and grocery store products with salmonella.

'How can we trust the Lib Dems when they don't know if they're yellow or orange?'
Ian Steadman, New Statesman, 2015

The official style guide for the British Liberal Democrat Party defines the primary colour for promotional materials as Pantone 1235 C. That shade, which some maintain is gold and others insist is amber, was selected in 2009 but has not been consistently applied, with an ever-changing array of orangey-yellows appearing as the backdrop for successive party conferences. The diehard Liberal Party, which refused to join the Liberal Democrats, stands by straightforward orange, which has been associated with the party led by Gladstone, Asquith and Lloyd George since the 1920s.

Orange is a logical choice for political parties that want to present themselves as progressive but not too left-wing – which partly explains why it has been adopted by the likes of the New Democratic Party (Canada), the Orange Democratic Movement (Kenya), Your Movement (Poland), the People's National Party (Jamaica) and the Social Democratic Party (Portugal).

Orange became a symbol of dissatisfaction with the Communist regime in Poland, when Waldemar Fydrych – a maverick humourist, art history student and draft-dodging 'major' – launched the Orange Alternative movement in Wrocław in 1981. As Ben Lewis recounts in *Hammer and Tickle*, his history of Communism through jokes: 'Fydrych's first action consisted

of graffitiing a symbol of an orange dwarf all over Warsaw. Solidarity activists had been painting their logo – an anchor – on walls across Poland. The authorities painted them over with white paint. Fydrych then added his orange dwarf on top of the splodges of white – an apt mockery of the government's efforts to suppress reform.'

Why orange and why dwarves? Part of the joke was that the colour combined the red of Socialism with the yellow on the Vatican flag.

Although the use of dwarves now seems politically incorrect, at the time, Fydrych told Lewis, they were chosen to expose the absurdity of Marxism–Leninism. The slogan 'Without dwarves there is no freedom' worked because, as he said, 'Can you take a police officer seriously when he is asking you: "Why did you participate in an illegal meeting of dwarves?"'

Between 1981 and 1991, the Orange Alternative organised more than sixty happenings across Poland. In one prank, worthy of the Situationist movement, protesters gathered around the chimpanzee cages at Wrocław Zoo and sung Stalinist songs. Alas, the Orange Alternative was crushed in Poland's first free presidential election in 1990, when Fydrych won just one per cent of the vote.

Major Fyrdych is not entirely without honour in his own country. In Wrocław, where he studied art history, 292 figurines of orange dwarves have become a tourist attraction. Characteristically, he has sued the council for breach of copyright.

In the winter of 2004, Fydyrch knitted an orange scarf to support Ukrainian protests against the fraudulent election of President Viktor Yanukovych and the poisoning of opposition leader Viktor Yushchenko. Fyrdych and his followers drove from Warsaw to Kiev in an orange battle bus and, in a ceremony witnessed by thousands of rebels, presented Yushchenko with a 15-metre-long orange scarf.

Our Ukraine, Yushchenko's centre-right party, had adopted orange as its official colour, to distinguish itself from the blue of Yanukovych and Communist red. When a well-organised youth movement occupied Maidan Nezalezhnosti (Independence Square) in Kiev to protest against the fixed election, they set up a

The Orange Revolution at its height, protesting the fraudlent election of Yanukovych, on Independence Square in Kiev, December 2004.

sprawling tent city bedecked in orange. Their courage inspired a revolution, which led to Yushchenko's victory in a second election.

The tremors from Ukraine's political earthquake were felt in Moscow. Interpreting 'orange fever' as an American-inspired, anti-Russian plot, Vladimir Putin's government co-opted ultra-nationalist street gangs to confront pro-democracy activists. State-sponsored anti-Orange protests were organised, which Russians were encouraged to attend by threats of dismissal or, in one instance, treats from an IKEA store. The movement's uncompromising ethos was illustrated by its official logo: an orange snake being strangled in a fist.

At the start of the 1990s, when Viktor Orbán made himself a national hero by declaring that the Soviet Union should withdraw its troops from Hungary, his Fidesz party adopted an orange circle as its official logo and called its weekly paper *Magyar narancs*

('Hungarian orange'). The title is intriguing because the country's Communist regime did try to cultivate oranges. Unsurprisingly, given the climate, they failed – a misbegotten enterprise parodied in Péter Bacsó's movie *The Witness* (1969). In one scene, a party leader, visiting a scientific cooperative, asks to taste the fruit of their labours. As the only ripe orange has been accidentally eaten, he is given a lemon, with the explanation: 'It's the new Hungarian orange, a bit more yellow, a bit more savoury but it's ours.'

Even for a creature as fearsome as the tiger, orange might not seem like the most practical colour to wear when you're out hunting. But it works, largely because the deer and other hoofed animals they prey on are colour-blind, so the orange of the tiger looks green, like the surrounding vegetation. To load the dice even further in the tiger's favour, their black vertical stripes break up their shape in a phenomenon which biologists call disruptive colouration – rather like dazzle camouflage (see *Grey*).

'As queer as a clockwork orange is an old Cockney slang phrase, implying a queerness or madness so extreme as to subvert nature, could any notion be more bizarre than that of a clockwork orange?' Anthony Burgess wrote in *The New Yorker* in 1973, explaining the title of his most controversial novel. Alas, 'clockwork orange' is not mentioned in dictionaries of Cockney slang. Burgess may have misheard, or embellished, the 1960s Cockney phrase 'as queer as a chocolate orange', which meant odd, unusual or homosexual.

PURPLE REIGN

I n Peter Paul Rubens' painting *Hercules and the Discovery of the Secret of Purple* (1636), the hero of Greek and Roman myth is worried that his dog's mouth is bleeding. Luckily, Hercules' faithful companion has merely bitten a snail. In Rubens' painting, the dog is pawing at a snail's shell, albeit completely the wrong kind of shell. This is how, according to the unreliable testimony of Greek rhetorician Julius Pollux, Tyrian purple was discovered.

So why isn't the dog's mouth purple? In truth, if you were a Phoenician, Roman or Minoan, you would probably say the mouth is indeed purple – Tyrian purple. In ancient times, this colour, manufactured from thousands of sea snails, varied from purple to dark red. Pliny the Elder observed: 'The

Tyrian colour is most appreciated when it is the colour of clotted blood, dark by reflected and brilliant by transmitted light.'

Making Tyrian purple was brutal, costly, complicated and smelly. In ancient times, the dye was usually extracted from three species of shellfish – the *Purpura* (*Bolinus brandaris*), the *buccinum* (*Stramonita haemostroma*) and *Murex trunculus* – which were found in abundance near the city of Tyre. (The dye was first created in Tyre, a Phoenician settlement, and it's possible that the Greeks named the Phoenicians after *phoinikes*, their word for Tyrian purple.) Caught in traps, the molluscs were taken ashore and cracked or squeezed open to extract a clear liquid from a gland, known as the 'flower' or 'bloom', at the head of the shellfish. When exposed to sunlight and air, this liquid would turn pale yellow-green, blue and then purple. Estimates vary, but it probably took around 250,000 shellfish to produce one ounce of the dye, which explains why it was so incredibly expensive: by the third century CE, Philip Ball writes in *Bright Earth* 'a pound of purple-dyed wool cost around three times the yearly salary of a baker'. The very best Tyrian purple cloth was called *dibapha*, meaning 'twice dipped'.

Tyrian purple became so deeply associated with Alexander the Great that, when he died in 323 BCE, a 'magnificent purple robe, embroidered with gold' was laid over him in his coffin, as his biographer Robin Lane Fox records. Among the prominent Romans who idolised Alexander were Julius Caesar and his adopted son, Octavian, who visited the conqueror's tomb in Alexandria. In 27 CE, after Octavian became the first emperor Augustus, Tyrian

Purple women: a Roman fresco preserved at the Villa of the House of Mysteries, Pompeii.

purple was restricted to the imperial family. The monopoly was enforced with varying degrees of enthusiasm by later emperors. Reputed to have commissioned his own purple bath, Nero (who reigned from 54 to 68 CE) was, Suetonius claims, so possessive about the imperial prerogative that when he spotted a woman at a recital wearing purple he 'pointed her out to his agents, who dragged her out and stripped her on the spot, not only of her garment, but also of her property'. That was almost lenient: the official line under Nero, and later Valentinian III (who reigned from 425 to 455 CE), was that any member of the lower classes caught wearing the imperial colour would be executed. During the reigns of Septimius Severus (193 to 211 CE) and Aurelian (270 to 275 CE), on the other hand, all women were permitted to wear purple, whereas under Diocletian (284 to 305 CE) anyone could wear the colour if they could afford to pay the tax on it. Always something of an outlier, Caligula (emperor from 37 to 41 CE) insisted that his beloved steed Incitatus only wear purple blankets.

In the Byzantine empire, the title *Porphyrogenitus* – 'born to the purple' – was reserved for a prince born after his father had become emperor and – in theory, at any rate – in the Purple Bedchamber at the Imperial Palace in Constantinople.

The traditional method of preparing Tyrian purple dye was lost to the West after Constantinople fell to the Ottomans in 1453. It was rediscovered in 1856 – coincidentally the year that William Perkin invented the synthetic dye mauve – when French zoologist Félix Joseph Henri de Lacaze-Duthiers spotted a Mediterranean fisherman drawing yellow designs on his shirt with the juice of a shellfish which turned purple-red in the sun.

In 1509, a new law gave King Henry VIII and his immediate family the exclusive right to wear purple or gold silk. The royal accounts reveal his penchant for purple velvet footstools and cushions. In 1542, he bought 'a gown of purple satin, furred with the sleeves and border set with 130 diamonds and 131 clusters of pearls, set in gold … and in every cluster is four green pearls'. When he was feeling generous, Henry allowed his hunting companions to wear purple, but such largesse could not be taken for granted.

Napoleon raised himself to the purple rather than being born into it, a point he made explicit on 2 December 1804, at his coronation in Notre-Dame cathedral in Paris. Wearing a purple velvet cloak, Napoleon picked up the imperial crown and placed

Napoleon on his Imperial Throne by Jean-Auguste-Dominique Ingres (1806).

it on his own head, to the chagrin of Pope Pius VII, who had travelled to Paris expecting to lend Bonaparte a hand. In Jean-Auguste-Dominique Ingres' portrait, *Napoleon I on his Imperial Throne* (1806), the new emperor is depicted wearing a purple cloak, a reference to the imperial glory of Augustus.

Banished to the island of Elba in April 1814, Napoleon assured followers that he would return bearing violets, a favourite flower from his Corsican childhood. As violets bloom in the spring, this was a cryptic message that he would stage his comeback by, at the latest, April 1815. Until then, Bonapartists could toast him as 'Caporal la Violette' without arousing suspicion.

In Japan, *murasaki*, a deep shade of purple, was for centuries *kin-jiki*, one of the colours forbidden to ordinary people. In 603 CE, Prince Shōtoku created a ranking of twelve levels of officials based on merit rather than birth (although most people at court would have been nobles anyway). Each level corresponded to key Confucian values – and was identified by a different coloured cap. Only ministers and administrators of the very highest rank – called *daitoku* or 'greater virtue' – were allowed to wear deep purple caps. The second highest rank wore a lighter purple.

At the time, purple dye was expensive in Japan, because it was based on the roots of the gromwell plant, which was also prized as a medicinal herb. The process itself was far from straight-forward. In 2015, in an experiment by dye expert Dr Kazuki Yamazaki and his students, it took repeated kneading, dyeing and setting with mordant over several days to produce a deep, complex purple in the traditional way.

Prince Shōtoku in deep purple murasaki.

In one popular tale, Prince Shōtoku gives food, drink and his purple cloak to a starving beggar, who is actually Bodhidharma, the monk reputed to have brought Zen Buddhism to China. Such tales enhanced purple's prestige, as did the use of purple and white in the costumes of emperors and gods in Noh, a classical form of dance-drama first performed in the fourteenth century. Even today in Japan, the colour is a mark of official distinction: a medal of honour with a purple ribbon is awarded for significant academic or artistic achievements.

❖ ❖ ❖

Kabuki actor Kuniya Sawamura performs *Sukeroku* in Tokyo, 2006.

The imperial hegemony over purple was eroded by another form of Japanese theatre, Kabuki, and one play in particular: *Sukeroku* (1713), better known in English as *The Flower of Edo*. The revenge drama's eponymous protagonist is, the American critic Jay Keister writes, a 'heroic, swaggering dandy … one of the most popular heroes of the Japanese stage'. Sukeroku's swagger was symbolised by his vivid make-up, his bullseye-patterned umbrella and, most famously of all, his purple headband (*murasaki hachimaki*). *Sukeroku* is still regularly performed in present-day Japan.

The role of Sukeroku soon became synonymous with Ichikawa Danjūrō, the stage name used by actors born in – or adopted into – the Ichikawa family. The actor Danjūrō II's purple headband was a gift from a high-ranking woman working for the shogun's mother, but he made it integral to the stylish, sexual and subversive dance with which Sukeroku makes his entrance.

The *Sasakia charonda* – also known as the great purple emperor – is the national butterfly of Japan. In 2019, a government study found that the population of this species, once found in abundance in forest canopies across the country, is declining so dramatically it could become extinct in the wild by 2025.

The word 'purple' comes from the Old English word *purpul* derived from the Latin *purpura* and, before that, *porphyra*, the Greek word for the Tyrian purple dye. The only word that fully rhymes with purple is 'curple', a corruption of the Scottish word crupper used to describe the rump of a horse or the strap that keeps a horse's saddle in place. Scotland's national poet Robert Burns rhymed 'curple' with 'proud imperial purple' in an admiring epistle to Mrs Scott, the 'guidwife of Wauchope House' in 1787.

Many shades of purple are associated with flowers and plants: violet, lavender, periwinkle, lilac, hyacinth, orchid, aubergine, boysenberry, mulberry, heather, heliotrope, orchil, thistle and phlox (aka psychedelic purple), which, for reasons too obvious to be credible, is often said to be Jimi Hendrix's favourite colour.

Food and drink have inspired a few more shades – notably grape, plum, raisin and sangria. Among the outliers are electric purple (what a hue halfway between violet and magenta looks like on your computer screen),; amethyst (paler than the gemstone it is named after) and 'Purple pizzazz', a lurid, pink-tinged Crayola crayon.

In *The Waste Land*, T.S. Eliot uses the many associations of purple – particularly in its violet, lilac and hyacinth shades – to suggest mystery, destruction, death, redemption and doomed romance. In the second line of the poem, he refers to lilacs emerging from the dead land, a likely allusion to Walt Whitman's lament for Abraham Lincoln ('When lilacs last in the dooryard bloom'd'). Lilacs traditionally signify the renewal of spring – they bloom in April – but by referencing the assassinated president, Eliot associates them with anguish, destruction and death.

Elsewhere in the poem, hyacinths are associated with failed love and untimely death. While the 'hyacinth girl' cherishes her visit to the 'Hyacinth garden' with her former lover, the memory perhaps stirs no such emotions in him. In Greek mythology, 'Hyacinth' is the name of a beautiful Spartan prince, killed by a stray discus, whose blood is turned, by his lover Apollo, into purple flowers engraved with 'Ai, ai', an ancient Greek cry of mourning. Though Eliot never tells us outright, from the allusions – and the association of purple hyacinths with regret, forgiveness and sorrow – it seems reasonable to assume that the flowers given to the 'hyacinth girl' were purple.

And there is the famous passage describing 'the violet hour', when late afternoon runs into early evening – a time of uncertainty and ennui (the typist passively succumbing to seduction), apocalyptic change and, ultimately, redemption. In the Christian church, purple is the liturgical colour of advent and lent, times of repentance and preparation for Christ's birth and resurrection. In the poem, this hour is profoundly violet, from the 'bats with baby faces in the violet light', to the atmospheric violence in the 'violet air' that renders such great cities as Jerusalem, Athens, Alexandria, Vienna and London unreal – a radical but necessary step if they are to be renewed.

The purple component of the official attire of British and American suffragettes – a white dress, with purple and green sashes (purple and gold in America) – was a celebration of female power. As Emmeline Pethick-Lawrence, editor of the suffragette newspaper *Votes for Women*, put it: 'Purple ... is the royal colour. It stands for the royal blood that flows in the veins of every suffragette, the instinct for freedom and dignity.' Shoes and underwear in suffragette colours were also available in Britain.

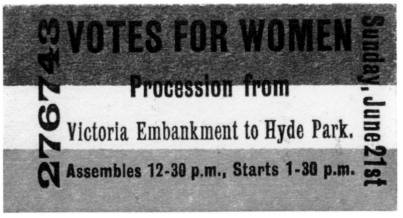

Admission ticket to the Hyde Park demonstration organised by the Women's Social and Political Union on Sunday 21 June 1908.

The liberating power of feminine purple was celebrated by English poet Jenny Joseph in her hugely popular 'Warning' (1961) with its opening lines 'When I am an old woman I shall wear purple. With a red hat which doesn't go.' Joseph, who died at the age of 85 in 2018, once admitted she never wore the colour, saying: 'I can't stand purple. It doesn't suit me.'

On 16 September 1660, Samuel Pepys wrote that he had gone to Whitehall, 'where I saw the king in purple mourning for his dead brother'. (The King was Charles II. His brother, the Duke of Gloucester, had died of smallpox.) In nineteenth-century Britain: widows whose husbands had died two years ago often wore the shade heliotrope in half-mourning. In some Catholic countries – especially Brazil – purple is associated with the Passion of Christ and is worn at funerals.

In Antigua, Guatemala, the association is reflected in the purple robes and conical hats worn by hundreds of penitents – known locally as *cucuruchos* or *capirotes* ('cones') – who at Easter carry a float commemorating Christ's suffering and resurrection. It takes seventy people to carry one float and fifty teams of penitents are needed for the procession, which is both a poignant religious ritual and a colourful, ephemeral living work of street art.

In 1856, a precocious 18-year-old scientist called William Perkin gave purple to the people. His experiments with aniline, a cheap extract from the waste product coal tar, to produce synthetic quinine to treat malaria had misfired, leaving a residue of dirty brown sludge in his beakers. As Perkin cleaned them with alcohol, the sludge turned a rich fuchsia purple. He had

accidentally discovered a synthetic dye. At first, he called this dye, quite inaccurately, Tyrian purple. It was renamed mauveine to make it sound trendier, but we now know it as mauve.

Luckily for Perkin, Empress Eugénie, wife of Napoleon III, decided that mauve matched her eyes (which, although lauded for their beauty, were definitely not purple). Her endorsement as a royal made the colour fashionable. In January 1858, Queen Victoria wore a rich mauve velvet dress at her eldest daughter, Princess Victoria's, wedding to Prince Frederick William (who was succeeded as Kaiser of Germany by their son, Wilhelm II).

Fashion was almost as fickle in the 1850s as it is today, and within a year of the royal wedding *Punch* magazine was bemoaning 'mauve measles'. The colour became associated with ageing ladies dressing younger than their years. In 1891, Oscar Wilde observed, in *The Picture of Dorian Gray*: 'Never trust a woman who wears mauve.'

The discovery that colour could be made synthetically and industrially was genuinely revolutionary – for the first time in history, ordinary people could afford to wear colours that previously had been the preserve of the wealthy and well-born. The same year Victoria promoted mauve, chemists in Germany had produced a second aniline dye, the bright red fuchsine, also known as magenta, which was cheaper to make than mauveine.

By the mid-1860s, synthetic colours had laid the foundations for the modern chemical industry and were revolutionising fashion. They did pose some challenges. William Morris complained that 'The fading of the new dyes is a change into all kinds of abominable and livid hues.' After leaving dyes in the sunshine to test how fast they were – and fast, in this sense, means resilient, as in 'hold fast', rather than quick – Scottish textile manufacturer James Morton (1867–1943) restricted his dyes to those that were, in his terms, 'sun dour', resistant to the sun. Labels such as Burberry then marketed these dyes – and clothes – as 'indelible'.

The greater challenge (which is a long way from being solved) is the indelible stain that fashion is leaving on our environment. The World Bank estimates that one fifth of global water pollution results from the dyeing and treatment of textiles. The sooner the industry switches over largely, or wholly, to non-toxic, biodegradable and durable dyes, the better for all of us.

We now know that the first synthetic colour, Egyptian blue, was invented in around 2200 BCE and used to paint statues, ceramics and the tombs of pharaohs. Roughly 1,400 years later, Chinese craftsmen created a synthetic colour known as Han purple. It is believed that Taoist glassmakers discovered it while perfecting jade-coloured glass. Making this colour was hardly straightforward – raw materials had to be ground in exact proportions and heated to around 850–1000 °C – but the Chinese produced enough Han purple (named after the dynasty under which it flourished) to dye beads, create murals and colour the sculptures depicting the Terracotta Army of Qin Shi Huang, the first emperor of a unified China. (Sadly, most of the purple on the emperor's funerary warriors oxidised after their excavation.)

Han purple – still a discernible tinge on the Terracotta Army.

The colour itself mysteriously disappeared, along with the Han dynasty, around 220 CE.

When the formula was recreated in 1992, physicists were astonished to discover that Han purple could collapse the third dimension. As Cassie Ryan wrote in *Vision Times*: 'When exposed to extreme cold and an intensely strong magnetic field, the pigment switches to a state called the quantum critical point where it "loses" its vertical dimension. This means that light waves travelling through can only move in two directions, possibly due to the mineral's tile-like structure.'

It took more than a thousand years for Christ's representatives on Earth to define a clerical uniform. In 1215, the Fourth Lateran Council forbade clerics from wearing red and green. The fall of Constantinople in 1453, which restricted supplies of Tyrian dyes in the West, prompted the Vatican to micro-manage clerical colours. In 1464, Pope Paul II decreed that cardinals should wear scarlet. Bishops were allocated a cheaper, pinker brand of purple, while the Pontiff monopolised 'true purple'. It was later suggested that red symbolised the cardinals' willingness to spill blood in defence of the church.

The Catholic church's colour codes became more elaborate over time, distinguishing between the ranks of the ecclesiastical hierarchy. The rules were simplified, up to a point, in 1966 by the Second Vatican Council. In the new spectrum, purple remained the preserve of the Pope and those dignitaries who assist in his ceremonial and civil duties, scarlet was allocated to cardinals (for their caps and sashes, and for their robes when in conclave) and black was designated for ordinary priests' cassocks.

In 1508, purple proffered tangible proof of papal hypocrisy. The sores that break out on the skin of syphilis patients were colloquially known as 'purple flowers'. Officially celibate, popes and cardinals should have been immune to this sexually transmitted disease, but on Good Friday in 1508, Pope Julius II had so many 'purple flowers' on his feet the faithful were not permitted to kiss them.

Purple is also the preserve of the two highest ranks of priest in Honmon Butsuryū-shū, a branch of Nichiren Buddhism. In Judaism, purple symbolises the union of feminine red (reflecting menstrual blood) and masculine blue, and the union of heaven (blue) and earth (red). If you believe in auras, a purple aura is associated with intuition, psychic powers and spirituality.

Bees can see a shade of purple that we can't – they have no photo-receptor for red, and their visible spectrum includes a combination

A bee's eye-view of colour. They can see their own kind of purple, a blend of yellow and ultraviolet light.

of yellow and ultraviolet light which scientists call 'bee's purple'. Using ultraviolet light to guide them to the pollen-rich parts of a plant, they can distinguish colours faster than any other species on the planet – five times quicker than we can – and are most attracted, studies conclude, to purple, violet and blue. As Sharla Riddle wrote in *Bee Culture*, the official magazine for American beekeepers: 'If bees were a superhero, sight would be their superpower.'

'I have finally discovered the true colour of the atmosphere. It's violet. Fresh air is violet.'
Claude Monet, in a letter to Paul Cézanne

On 5 December 1926, the undertaker was about to cover Claude Monet's coffin with a black cloth, when Georges Clemenceau, the former French prime minister and a good friend of the deceased, intervened. Remembering that the artist had sought to eliminate black from his work, Clemenceau, muttered, 'No! No black for Monet!' A flower-patterned violet cloth from Monet's home in the Normandy village of Giverny was used instead.

The Impressionists loved purple so much that critics accused them of 'violettomania'. (Kandinsky dismissed the shade as 'morbid'.) Such complaints did not bother Monet, who wrote to Paul Cézanne in 1869: 'Somewhere between violet and green, that's where the colour lies that connects everything, there, somewhere in infinity, between the tints of air and water.' Monet's paintings of water lilies – there are approximately 250 of them – embody this credo.

Cézanne once said of Monet, 'only an eye but good lord what an eye'. Critics sneered that Monet's eye was only unique because his optic nerves were weak and his eyeballs trembled, symptoms usually associated, at the time, with hysterics and lunatics.

From 1908 onwards, Monet did struggle with his sight, but did it affect the way he saw – and painted – colours? In 1912, he was diagnosed as having cataracts in both eyes, and he began to label his tubes of paint, and place the paints in the same order on his palette, to avoid choosing the wrong colour. As Anna Gruener noted in her 2015 essay on Monet's cataracts in the *British Journal of General Practice*: 'His brush strokes became broader and his painting, like his cataracts, more brunescent [brown and opaque].' In 1923, the cataracts – and the lens of his right eye – were surgically removed and eventually his sight stabilised when the Swiss company Carl Zeiss supplied him with new optical lenses.

Did the removal of the lens from his right eye enable Monet to see ultraviolet light? Normally, our eye lenses screen out

Monet's Waterliliies – set against 'the colour that connects everything'.

ultraviolet light – we are not really sure why, but they do – but some people with aphakia, who are born without a lens, or lose one through injury or surgery, insist they can see it. American academic Bill Stark, who is aphakic in one eye, says ultraviolet appears to him as whitish blue or whitish violet. Others have said it gives everything a blue cast or that objects appear reddish. Both these colours were prominent in Monet's paintings after the operation and arguably more prominent than violet. Devoid of any sharp physical detail, those final works look like early iterations of Abstract Expressionism.

Purple graces the flags of just two countries: Nicaragua (it appears in the rainbow in the centre) and the republic of Dominica (in the form of the purple-chested sisserou parrot, an endangered species found in the wild only in this Caribbean country). As a symbol of communal togetherness, the colour also features on the official asexual pride flag.

'Parent alert! He is purple, the gay pride colour, and his antenna is shaped like a triangle – the gay symbol.'
An article in the February 1999 edition of Jerry Falwell's *National Liberty Journal*

American evangelist Jerry Falwell was pilloried for homophobia after an article in his *National Liberty Journal* claimed that Tinky Winky, the purple Teletubby, was promoting homosexuality. Suspicions had allegedly been aroused because, in addition to being purple, the character liked big hugs, carried a red bag and sang 'Pinkle winkle, Tinky Winky'. Falwell – who wasn't the author of the article – said he had never watched *Teletubbies* and insisted that he didn't know whether Tinky Winky was homosexual or not.

And it wasn't Falwell's magazine that first 'outed' Tinky Winky. Online speculation about the purple Teletubby's sexuality began soon after the series was first shown on the BBC in 1997. In January 1999, the *Washington Post*, that paragon of liberal virtues, hailed 'Tinky Winky, the gay Teletubby' as the 'new Ellen DeGeneres'.

Even before Tinky Winky became a minor camp icon, purple – particularly in the shade of lavender – had been associated with gays. Marc Almond acknowledged as much in his 2010 song 'Lavender': recalling British society in the 1950s and 1960s, a

MARC ALMOND
TRIALS *of* EYELINER

time when homosexuality was illegal, clandestine and dangerous, Almond hails 'St Dirk of Bogarde' and repeats the refrain, 'He's got a touch of lavender'. Though Bogarde never publicly acknowledged his sexuality, to avoid jeopardising his acting career, he did give an extraordinary – and risky – performance as a crusading barrister resisting his feelings for a young construction worker in Basil Dearden's blackmail drama *Victim* (1961), reputed to be the first British movie to mention the word 'homosexual'.

In the 1950s, 'lavender' was used as a pejorative term to persecute gays and lesbians working for the American government. Senator Joseph McCarthy (who once declared that anybody who opposed him was 'either a Communist or a cocksucker') attacked Harry Truman's administration with such venom and vehemence that the State Department 'allowed' ninety-one gay employees to resign.

Before the 1952 presidential election, Republican Senator Everett Dirksen had vowed to kick the 'lavender lads' out of the diplomatic corps and, on 27 April 1953, Republican President Dwight Eisenhower signed Executive Order 10450 banning gay men and lesbians from working for the federal government. The official excuse was that they posed a blackmail risk. As many as 5,000 employees lost their jobs. In the deeply conformist America of that era, the stigma of being 'outed' was

Hamish Bowles attends New York's 2019 Met Gala 'Celebrating Camp'.

too much for some, who committed suicide. Largely nullified by the courts, Eisenhower's ban applied to the military until it was rescinded by Bill Clinton in 1995.

The 'lavender lads' scandal may have influenced Falwell, who was twelve when McCarthy's witch-hunts began. His animus certainly ran deep. In 1981, he wrote: 'Please remember, that homosexuals do not reproduce. They recruit! And many of them are after my children and your children.' The inescapable conclusion is that Falwell did believe in a gay conspiracy – he just didn't think it was led by a purple Teletubby.

The part of New Orleans' French Quarter that is home to the Café Lafitte in Exile, founded in 1933 and reputed to be America's longest continuously operating gay bar, is known as the Lavender Line. Britain's first lesbian, gay, bisexual and transgender telephone helpline, which opened in Brighton in 1975, had the same name.

'When there's blood in the sky, red and blue … purple … purple rain pertains to the end of the world and being with the one you love and letting your faith/god guide you through the purple rain'.

Prince, on the inspiration for his greatest hit

Prince Roger Nelson, to use his full name, had a penchant for purple jumpsuits, and the music room of his home and studio in Paisley Park, Minnesota, was thoroughly purple. *Purple Rain* was his most famous song, album and movie. The Minnesota Vikings, his favourite American football team, wear purple shirts and were honoured in his song 'Purple and Gold'. (In the 1970s, the Vikings' watertight defence was known as the 'purple people eaters' after the novelty hit single.) Prince's reputed devotion led Pantone to posthumously award him his own purple, Love Symbol #2. Yet in August 2017, Prince's sister Tyka Nelson told the *Evening Standard*: 'It is strange because people always associate the colour purple with Prince, but his favourite colour was actually orange.' It's hard to reconcile her remark with a rock star who lived in

a world where, to quote rock journalist Ian Penman, 'Every purple detail (no matter how small) is interlinked with every other, from top to bottom, in his unique control-freakish way'. Even when Prince posed naked on the cover of his 1998 album *LoveSexy*, his head was surrounded by a nimbus of purple petals.

Prince isn't the only rock star with a passion for purple. The rock band Deep Purple were named after the eponymous Bing Crosby song, a favourite of Ritchie Blackmore's grandmother. Jimi Hendrix's 'Purple Haze', often interpreted as a song about psychedelic drugs, may have been inspired by his love of science fiction. In 1966, Hendrix read – and dreamed about – Philip José Farmer's novel *Night of Light*, in which a 'purplish haze' unsettles inhabitants on a distant planet. He also said that the song had been inspired by another dream, in which a woman in New York was trying to romance him with voodoo (which would certainly tally with the line 'that girl put a spell on me'). There probably is no single meaning. Hendrix told an interviewer once: 'Ooh, you should hear the real "Purple Haze". It has about ten verses, but it goes into different changes. It isn't just "Purple Haze, blah, blah blah".'

Emperors and rock stars are hardly the only people to get possessive about purple. Having selected a particular shade to evoke the luxurious taste of its chocolate, Cadbury waged an exhaustive, expensive seven-year legal battle against Nestlé to protect its monopoly, even going so far as to harry people at Church of England fairs who sold purple-wrapped fair-trade chocolates bearing a Christian message. Cadbury finally gave up after the Court of Appeal decided, as one legal expert put it, 'it didn't want to give anyone a monopoly on purple'.

The American Supreme Court has taken a different view, concluding, in 2017, that colour could be registered as a trademark, providing there was evidence to prove that it had become associated with a particular product and implied that the product came from a particular source.

Purple is worn by Prince's beloved Minnesota Vikings and a few soccer teams, but it is odd, given how visible it is against green grass, that it is not more popular in sports. Where purple is chosen, it is often for very specific reasons: Perth Glory adopted it to symbolise the rebirth of soccer in Western Australia, because it was not one of the game's traditional colours. In Japan, Kyoto Sanga's purple shirts reflected the city's imperial past and association with the Buddhist priesthood.

Anderlecht, Austria Wien and Real Valladolid all play in purple, but the most celebrated of purple-wearing football squads is Florence's team, Fiorentina, nicknamed *La Viola* ('The Purple'). When the club was founded, in 1926, the shirts were half red and half white. Club legend has it that the switch to purple, two years later, came about when someone made the mistake of washing the kit in the Arno River, accidentally turning the shirts the colour the team has worn ever since. In 1961, Leeds United adopted an all-white kit because new manager Don Revie aimed to emulate the success of Real Madrid, who had won the first five European Cups in the same colour. The second of Real Madrid's European Cup triumphs – in 1957 – came against Fiorentina. If *La Viola* had prevailed, would Revie have been tempted to choose purple?

'That is a deep violet, please, depend on it! Not so rose'

Franz Liszt instructing his orchestra – the musicians thought he was joking until they realised that he saw sounds as colours.

Nobody had ever seen a harpsichord like it. Louis-Bertrand Castel (1688–1757), a French Jesuit, mathematician and physicist, had put a six-foot square frame on top of a harpsichord, containing 60 differently coloured glass windows, each backed by a lighted candle and fronted by a small curtain attached to a specific key on the harpsichord. When he struck a key, the curtain in front of the appropriate pane was raised to reveal a colour.

Castel's attempt to synchronise the colour spectrum and the musical scale on his *clavecin oculaire* (light keyboard) led many curious souls – including German composer Georg Philipp Telemann – to travel to Paris to see him play it in his studio. Castel dreamed that one day every household in Paris would have its own *clavecin oculaire*, but his invention nearly bankrupted him. The instrument would not have been cheap to buy or run (think of all those candles). One scheduled performance, at London's Soho Square concert rooms, was cancelled as a fire hazard. None of Castel's instruments or diagrams have survived.

His quest had been inspired by Jesuit polymath Athanasius Kircher's observation: 'If, when a musical instrument sounds, someone could perceive the finest movements of the air, he certainly would see nothing but a painting with an extraordinary variety of colours.' Castel's experiments began with a colour keyboard based on Newton's ROYGBIV (Red, Orange, Yellow, Green, Blue, Indigo and Violet) spectrum. After concluding that there were at least a thousand colours in the rainbow, Castel created a new keyboard that matched twelve colours – crimson, red, orange, faun, yellow, olive, green, celadon, blue, agate, dark violet and violet – to the twelve tones of the musical scale. In his scheme, the note A was violet and B was dark violet.

A caricature by Charles Germain de Saint Aubin of Louis-Bertrand Castel's colour organ, proposed by the Jesuit mathematician in 1725.

As James Peel wrote in his essay 'The Scale and the Spectrum', Castel thought of colour-music as akin to the lost language of paradise, when all men spoke alike. He claimed that, thanks to his instrument's ability to paint sound, 'even the deaf listener could enjoy music'. At the climax of Steven Spielberg's sci-fi blockbuster *Close Encounters of the Third Kind* (1977), humanity and aliens learn to communicate in a universal language, combining music and colour. It makes you wonder what Castel might have accomplished if he had lived in the age of the computer.

One of Castel's kindred spirits was mystical Russian composer Alexander Scriabin (1872–1915), who used a chromola, a colour organ designed by American lighting specialist Preston Millar, in his twenty-minute symphonic extravaganza *Prometheus: The Poem of Fire*. Effectively a multimedia show before the term had even been invented, *Prometheus* should, to fulfil Scriabin's vision, have been performed in auditoriums bathed in the colours that corresponded to his music. In his system, red was the colour of Abaddon, angel of destruction and king of the locusts in the Book of Revelation, while blue and violet stood for reason and spirituality.

As a passionate believer in Theosophy, a mystical blend of philosophy and religion much in vogue at the time, Scriabin believed that his system represented a universal truth. His association of E flat with 'steel colour with metallic sheen', for example, may strike some as risible, but his conviction that there was a mystery linking music, poetry, dance and art would inspire others, notably the Australian artist, colour theorist and viola player Roy de Maistre.

Written in 1910, *Prometheus* was never performed to Scriabin's satisfaction. He often played a transcription of it at home, with his second wife Tatiana Schlözer working the lights. He came to believe that he was God, destined to summon people to the foothills of the Himalayas to synthesise their senses and the arts with his piano playing, perfume and poetry. A week before the composer's death,

a performance of *Prometheus* at New York's Carnegie Hall, on 20 March 1915, prompted *Scientific American* magazine to hail colour-music as a new art form. But the concert was not true to Scriabin's vision. The colours did not flood the auditorium but were projected onto a screen. One reviewer derided it as a 'pretty poppy show'.

In Alice Walker's career-defining novel *The Color Purple*, blues singer Shug Avery observes that 'I think it pisses God off when you walk past the colour in a field and don't notice it.' Shug is encouraging the novel's heroine Celie to enjoy the good things in life (including sex) that God has provided. The choice of colour also has a racial dimension. Walker has since said that purple represents 'womanism' – a term she uses for the liberation of African-American women – as opposed to 'feminism', a term for white American women which she equates with the colour lavender. In the novel, purple – actually aubergine – is the colour of the bruises suffered by Celie's friend Sofia, who is badly beaten by the police. Celie makes Sofia a pair of red and purple trousers, symbolising the colour's redemptive power.

'The General, ever desirous to cherish virtuous ambition in his soldiers, as well as to foster and encourage every species of military merit, directs that whenever any singularly meritorious action is performed, the author of it shall be permitted to wear on his facings over the left breast, the figure of a heart in purple cloth, or silk, edged with narrow lace or binding.'

George Washington, creating the Badge of Military Merit, 1782.

When the Badge of Military Merit – the precursor of the Purple Heart – was created, America had all but won its War of Independence. Peace talks had been under way for four months in Paris. Unusually for the time, the new honour was open to non-commissioned officers and soldiers, reflecting Washington's conviction that 'the road to glory in a patriot army and a free country is thus open to all'. He probably only presented the award to three or four soldiers, all sergeants. The war officially ended on 3 September 1783 and, although the badge was never actually abolished, it hung in abeyance

THE UNITED STATES OF AMERICA

TO ALL WHO SHALL SEE THESE PRESENTS, GREETING:

THIS IS TO CERTIFY THAT
THE PRESIDENT OF THE UNITED STATES OF AMERICA
HAS AWARDED THE

PURPLE HEART

ESTABLISHED BY GENERAL GEORGE WASHINGTON
AT NEWBURGH, NEW YORK, AUGUST 7, 1782
TO

until February 1932, the bicentennial of Washington's birth, when the War Department announced the Order of the Purple Heart. Official estimates suggest that more than 1.8 million of these medals have been awarded.

But why purple? The colour would have stood out on the brown, blue and grey uniforms worn by American soldiers during the War of Independence. It is often stated as an incontestable fact that purple symbolises courage in American culture, yet there is little conclusive evidence of this association before the 'figure of a heart in purple cloth' was instituted by Washington in 1782.

We know that America's commander-in-chief, and first president, was very particular about his colours – he complained bitterly when the verdigris green in his dining room at Mount Vernon did not turn out as expected – so the association with purple may have been very personal to him. Describing Tyrian purple, Pliny calls it 'the badge of noble youth'. In Washington's mind, the association may go back to one particular noble youth: Alexander the Great. He consulted Roman writer Quintus Curtius Rufus's biography of the conqueror at critical moments in his military career. He also owned a set of five oversized prints depicting Alexander's triumphs. One of them was entitled *Virtue surmounts all difficulties*. Alexandrian purple may have struck Washington as an appropriate colour with which to reward the 'virtuous ambitions' of America's soldiers.

SEA OF GREEN

Most authorities agree that green is one of the world's favourite colours – depending on which survey you believe, it is either second to blue or third behind red. We are hardwired to like green. Life on earth is maintained by the process of photosynthesis, and the representative colour of photosynthesis is the green of chlorophyll. Green spaces are good for our bodies – and our souls. In the words of Alistair Griffiths, Director of Science and Collections at the Royal Horticultural Society, 'We know that exposure to nature reduces stress and anxiety, improves mood and lowers blood pressure and heart rate. A 2002 study of 3,000 Japanese citizens aged 74 and over found that access to green walkable paths and areas increased their life expectancy – and studies of older people in the UK have shown similar results. You could call it the power of Vitamin G.' But green can represent bad luck, as well as good, and has connotations with gangrenous decay as well as fertility and rebirth.

❖ ❖ ❖

The Green Man, who has given his name to so many British pubs, was probably a descendant of such gods of fertility as Bacchus, Dionysus, Dumuzi (a Mesopotamian deity also known as Tammuz) and Osiris. Adopted – and adapted – as a symbol of regeneration by the early Christian church, he tends to be depicted surrounded by greenery, or with a face comprised, wholly or partially, of vegetation. A particularly impressive example of the latter can be found at Bamberg Cathedral in Germany, where the Green Man's head is shaped like an Acanthus leaf. If you look hard enough you can spot more than sixty Green Men at Winchester Cathedral – and, quite possibly, two or three Green Women. There are estimated to be seventy of them at Chartres. (The frequency with which images of the Green Man have been found in England and France suggests that the figure was especially resonant in Celtic society.) Although Michelangelo used the motif in the decoration of the Medici Chapel in Florence, the Green Man had lost his symbolic resonance by then, becoming little more than an adornment.

❖ ❖ ❖

The Green Knight, the central figure in the fourteenth-century poem *Sir Gawain and the Green Knight*, is a figure so baffling that J.R.R. Tolkien, who translated it into modern English, described him as 'the most difficult character to interpret'. The fact that the knight's skin and clothes are green would seem an obvious allusion to the Green Man, but he has also been associated with the Devil (Gawain regards the Green Chapel, where he is to meet the knight, as a hellish place), Jesus (green being the colour of resurrection) and even al-Khidr ('the green one'), a prophet, visionary and magus who can magically make arid land fertile. Al-Khidr is mentioned twice in the Hadith, the collection of Muhammad's sayings, and he features (albeit not by name) in the Qur'an. It's possible that his legend spread to Europe after the Crusades. Gawain's supernatural antagonist may also have been

A painting from the original manuscript of *Sir Gawain and the Green Knight*. The Green Knight is holding up his severed head.

inspired by Bricriu's Feast (*Fled Bricrenn*), a mythical Irish tale dating from the eighth century, in which the hero Cúchulainn beheads a giant in a green cloak, just as Gawain does.

One of the Green Knight's scariest distant relatives is the Green Hunter (*Der grüne Jäger*), one of the strangest fruits of the medieval imagination. Accompanied by the living, the dead and sundry malevolent creatures, the hunter is condemned to chase his prey through the night without ever catching it, a fate reminiscent of that eternal spectral navigator, the Flying Dutchman. If you encountered him, you had a good chance of being carried off to your death. This baleful legend is echoed in Goethe's most famous ballad, 'Der Erlkönig' ('The Alder King').

In 1964, Major General Peter Young, commander of the British Army in Cyprus, took out a blunt chinagraph pencil and drew a green line on a map to separate the the island's warring ethnic communities, who had recently agreed a ceasefire. The north was assigned to Turkish Cypriots, the south to Greek Cypriots, and the buffer zone Young had drawn between the two became known as the Green Line. The line did not prevent civil war in 1974, but it is still there, stretching for more than 300 kilometres and patrolled by UN troops.

In April 1975, Beirut got its own Green Line, improvised rather than ordained, marking the beginning of a brutal fifteen-year civil war. This Green Line, which mostly followed the course of Damascus Street, starting at Martyrs' Square in the north and running roughly eight kilometres south, divided mainly Christian East Beirut from mainly Muslim West Beirut. Over this barrier

Beirut's Green Line, slowly returning to nature, in 1990.

of earthen mounds, sandbags, barbed wire, overturned buses and shelled-out buildings, snipers from both factions shot at each other. Although people still lived on and around the street – and could, if they were prepared to accept the risk, cross it at certain checkpoints – shrubs, bushes, plants and trees began to reclaim the buildings and the road.

Given its obvious associations with nature, green emerged curiously late as the defining colour of environmental action. In the 1970s, the yellow and red of 'Nuclear Power? No Thanks' was more visible. The arrival of 'environmental green' can be dated to a time and place: September 1971, when Canadian activists chartered a ship to protest against nuclear tests planned for Amchitka, one of the Aleutian Islands off the coast of Alaska. They nicknamed the ship 'Greenpeace' and, although they were turned back by the US Navy, the name won public attention. When the activists – who had formed as the 'Don't Make a Wave Committee' – launched a campaign group in 1972, the Greenpeace name stuck.

Political parties, however, still lagged behind. The United Tasmania Group, believed to be the first political party to fight elections on a green manifesto, adopted a black, white and red logo when campaigning in Australia in 1972. Britain's Ecology Party, formed in 1975, had a green logo, but took a decade to rebrand themselves as the Green Party, prompted by the electoral success of Die Grünen, launched in West Germany in 1980. The German Greens adopted a sunflower symbol, a design that the oil and gas giant BP appeared to co-opt when it redesigned its traditional shield as a 'sunburst' in 2000. This attempt at greenwashing – BP were dabbling with solar investments under the slogan 'Beyond Petroleum' – foundered when the company spilled 134 million gallons of oil into the Gulf of Mexico in the Deepwater Horizon disaster in 2010.

❖ ❖ ❖

Since 1968, the principle that red means 'Stop' and green means 'Go' has been codified in the Vienna Convention on Road Signs and Signals. In Japan, however, green lights are more blue than green. The National Police Agency's traffic manual advises pedestrians that when a light turns *ao* – originally a 'grue' word which once signified green but now usually means blue – they can cross the street. The English translation tells pedestrians to cross on green. As Peter Backhaus noted in the *Japan Times* in 2019, 'Both texts refer to the same illustration: the image of a traffic light in red, yellow and, well, something turquoise.' The discrepancy arose in 1973, when the Japanese government decided, rather than to change *ao* to the unambiguously green *midori* in its traffic regulations, to modify the lights instead, making them a bluish hue of green.

The world's only known upside-down traffic light, sited on Tipperary Hill, in the New York town of Syracuse, reflects a dispute over the hidden meanings of red and green. It is green at the top, yellow in the middle and red at the bottom because some Irish-American residents kept throwing stones at the original lights (installed in 1925), incensed by the idea that the colour sequence, with 'British' red at the top, implied superiority over 'Irish' green at the bottom. After three years of repeated vandalism, city officials finally swapped the positions of the red and green lights.

San Francisco surgeon Harry Sherman created the world's first green operating theatre in 1914. Finding that the glare from the white walls, sheets and uniforms made it difficult to discern the finer anatomic detail of the patients he was operating on, Sherman maintained that 'spinach green' would act as a better complement to blood red, and persuaded the authorities at

Harry Sherman's ideas of spinach green to calm his patients was modified by colour consultant Faber Birren.

St Luke's Hospital to let him turn the walls, sheets and surgeons' uniforms that colour.

At around the same time, on the opposite coast of the USA, architect William Ludlow, deploring the 'white sterility' of hospital interiors, advocated the use of 'calming green' and other natural colours. The belief in green's therapeutic powers peaked in the 1930s when colour consultant Faber Birren adapted Sherman's 'spinach green', promoting a paler, milder green which was 'slightly passive in quality', to help doctors and nurses focus on the job. (Bright colours, he argued, 'tend to stimulate an outward attention towards the environment'.)

Hospitals are now significantly less green than they were. Surgical scrubs are often blue because they also refresh the surgeon's vision and make it easier for them to interpret a patient's bloody innards.

Not all blood is red. In New Guinea and the Solomon Islands there's a genus of skink (a kind of lizard), that has toxic lime green blood. Odder still, they are so stewed in a bile pigment called biliverdin that their bones, muscles and tissues are green, too. We don't conclusively know why these skinks have evolved in this way but the current hypothesis is that it protects them against some parasites, especially those that spread malaria.

The association between green and decomposition has been cemented by the word 'gangrene', which actually, despite the way it sounds, has nothing to do with the colour at all. The word comes, via Latin, from the ancient Greek *gangraina*, which signifies putrefaction.

❖ ❖ ❖

America's driving instructors are often phoned by prospective pupils seeking reassurance that they will not be learning in an 'unlucky' green car. There is no evidence of such questions being asked about any other colour. Nor is there any scientific evidence that green – or any other colour – causes more accidents (although, globally, fewer than one in fifty cars are green). And it's perplexing that the colour we associate with the life-giving qualities of nature is often considered to bring bad luck.

In the world of American motor racing, green's notoriety can be traced back to 17 September 1911, when, during a race in Syracuse, New York, Lee Oldfield's green Knox racing car blew a tyre and ploughed into the crowd, killing nine spectators and injuring fpirteen, in what is still the deadliest accident in the history of American motor racing. Oldfield escaped with minor injuries. Nine years later, the drivers were not as lucky. On 25 November 1920, at the Beverly Hills track, a green Fontenac driven by Gaston Chevrolet, the brother of the founder of the eponymous car company, crashed into Eddie O'Donnell's Duesenberg. Both drivers died, as did Lyal Jolls, O'Donnell's mechanic.

Many American racing drivers consider green an unnecessary risk. Four-time Indy 500 winner Rick Mears had the green wires in his cars repainted red, and Tim Richmond refused to drive a car sponsored by Folgers decaffeinated coffee because their brand was green.

The superstition about green has no such sway in British motor racing. Racing green has been associated with the country's motorsports since 1902, when Selwyn Edge won the Gordon Bennett Cup Grand Prix in an olive-green Napier car. (The colour reflected the fact that other countries, which had been racing for longer, had already laid claim to red, blue and white.) He was the only driver to finish the race, but even so, as the champion, he had the right to stage the next Grand Prix in England. Because motor

Jim Clark in his Lotus 25-Climax during the 1964 Monaco Grand Prix.

racing was still illegal in Britain in 1903, the race was relocated to Ireland and the Napier vehicles were painted shamrock green to honour the host nation.

There has never been a definitive shade called 'Racing green' but most of the paints sold under that name are much darker than the original shamrock green. It was in these more sombre shades that Jim Clark, Jack Brabham and Denny Hulme won the Formula One title with British teams in the 1960s. But on 7 April 1968, driving a green and yellow Lotus 48 at Germany's Hockenheimring, Clark veered off the track into the trees. He died on the way to hospital, aged just 32.

Green's racing pedigree began with charioteering. In the days of the Roman republic, there were four colour-coded teams – blue, green, red and white – but by the sixth century CE, after the collapse of the western Roman empire, the races in Constantinople, capital of the Byzantine empire, were being contested solely by green and blue teams. It's widely thought that the Blues were broadly perceived as representing the ruling classes, while the Greens stood for the masses, but, whatever the affiliations might have been, conflicts between the two camps were sometimes murderous. In 501 CE, for example, the Greens massacred 5,000 Blues in an ambush in Constantinople's amphitheatre. Supporters of both colours united against the emperor Justinian in 532 CE, in an ill-fated revolt which led to 30,000 deaths and signalled the end of charioteering as a popular sport.

Lincoln green, the colour worn legendarily by Robin Hood – and, rather fetchingly, by Errol Flynn in the Technicolor swashbuckler *The Adventures of Robin Hood* (1938) – was one of two textile colours closely linked to that city in the Middle Ages, the other being Lincoln scarlet. To make Lincoln green, the wool was dyed blue with woad and dyed again with a yellow plant called 'dyer's broom', creating an olive green that was somewhat more reliable than the other widespread green dye of at period – Kendal green. The modern shades of Lincoln green are different: the Pantone colour – hex #195905 – is much darker than the exhilarating Lincoln green worn by Errol Flynn. Dulux's Lincoln green sits about halfway between the two.

The rebellious, freedom-fighting outlaw who stole from the rich and gave to the poor is largely derived from Sir Walter Scott's novel *Ivanhoe* (1819), but it's likely that the fantasy Robin Hood was inspired by at least one real person. Historical records suggest that Roger Godberd is a plausible candidate. Having supported Simon de Montfort and the barons against King Henry III, he lived as an outlaw in Sherwood Forest, with a gang of as many as a hundred supporters, before being caught in 1272 and subsequently pardoned. According to the warrant issued for his arrest, he 'carried out so many great homicides and robberies that no one could pass through ... without being taken or spoiled of his goods'. It would have made perfect sense for Godberd and his men to camouflage themselves in Lincoln green, not least because foresters traditionally wore green.

It also made financial sense, costing half as much as Lincoln scarlet, which was made using a substance called kermes, after the insects that were crushed to make it. Kermes had to be imported from Turkey, a process that could be costly, lengthy and hazardous. As the more expensive cloth, scarlet signified a certain social standing and affluence. In one eighteenth-century ballad, Robin

Errol Flynn models Lincoln green, with matching sword.

Hood wears green in the forest and scarlet at court. The name of his nephew, Will Scarlet, mentioned in the ballad 'A Gest Of Robyn Hode', may reflect the young outlaw's love of fine clothes. As some songs and plays were commissioned by guilds, it could equally have been a bit of product placement, ensuring that both of Lincoln's most popular shades of textile would be promoted whenever the legend was recounted.

Do not wear green at sea, get on a green boat or have your boat painted green if either you or your fellow sailors are even slightly superstitious. Fishermen in various parts of England and France fear the colour will attract thunder and lightning. Other sailors, believing that green is the colour of land, worry that a boat painted that colour is likely to run aground. There is also a legend that green is shunned because it's associated with the corpses of officers and crew who have died on board.

Every 17 March, rivers and beer across the world turn green to celebrate Saint Patrick's Day, to celebrate the missionary and bishop who is credited with founding Christianity in Ireland in the fifth century. Yet the earliest known image of Saint Patrick, dating back to the thirteenth century, shows him wearing blue robes, not green, and in early Irish mythology the country's sovereignty – Flaitheas Eireann – was often represented by a woman in a blue robe. When Henry VIII declared himself king of Ireland in 1541, he gave the country a coat of arms: a golden harp on a blue background (which lives on today in the flag of the Irish president). Gradually, as divisions between the Irish people and the English crown deepened, blue became a tainted colour. By

The Chicago River turns green for Saint Patrick's Day.

the seventeenth century, the green sprig of shamrock had become an emblem of Saint Patrick, and it soon evolved into a motif of Irish nationalism.

In Russia, where alcohol is responsible for about 30 per cent of deaths (according to the World Health Organization), the victim of a fatal drinking binge is described as having been 'killed by the green snake'.

Traditionally, at weddings in the Scottish Lowlands, green could not be worn or used in decoration. The superstition was so strong that green vegetables were never served at the meal. The old English rhyme 'Married in green/Ashamed to be seen' may reflect the belief that the colour is associated with fairies, who – so the superstition goes – are inclined to cause mischief on wedding days.

William Shakespeare helped make green synonymous with jealousy. He first used the image in *The Merchant of Venice* (c.1596), when Portia talks of 'green-eyed jealousy' in love, then returned to it in *Othello* (c.1604), whose tragic hero is undone, as his nemesis Iago says, by the 'green-eyed monster that doth mock the meat it feeds on'. Where did this association come from? As far as we can tell, the most likely source is the Bard's own imagination.

In China, the expression 'wearing a green hat' signifies that a man's wife or girlfriend has been unfaithful. There are different explanations for this – some say the phrase sounds like 'cuckold' in Mandarin, others that prostitutes once wore green. In 2015, when Shenzhen police gave jaywalkers the choice of wearing a green hat or paying a $3 fine, one in three offenders paid up.

It was the custom in seventeenth-century France to carry bankrupts to the marketplace so that their disgrace could be publicly announced. The only way the offender could stay out of jail was to wear a green bonnet.

We tend to associate Satan with the colour red, but in the late Middle Ages his perfidy was often green. In Michael Pacher's *The Devil Holding Up the Book of Vices to St Augustine* (1471–75), the

Devil is painted in slimy green and looks like a prototype of the green aliens that pervaded so much twentieth-century pulp sci-fi. Noting that green devils began appearing in Europe's churches as early as the twelfth century, Michel Pastoureau speculates that the antipathy reflects the use of green by the armies of Islam.

The association between colour and the Devil is a motif in Chaucer's 'The Friar's Tale', in which the yeoman, sporting a short green jacket, admits to making a living out of extortion, trickery and violence, before telling his friend, the Summoner:

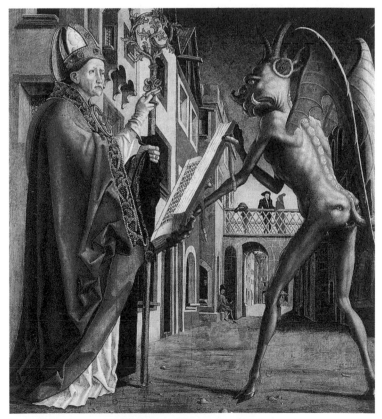

The Devil wears green in Michel Pacher's Kirchenväter altarpiece.

Eighteenth-century engraving of a Jack in the Green, possibly by Cruikshank.

'I am a feend, my dwelling is in helle.' Many of the creatures associated with Satan – dragons, witches and basilisks – were often portrayed as wholly or partly green. And green became the default colour for witches in popular culture after Margaret Hamilton's turn as the Wicked Witch of the West in MGM's psychedelic adaptation of *The Wizard of Oz*. In L. Frank Baum's novel, she is white.

'Jack of the green had made his garland by five in the morning and got under his shady building by seven'
Morning Chronicle and Advertiser, 2 May 1775.

Greenery has had a starring role in May Day rituals since medieval times, but it became the custom, in eighteenth-century Britain, for young chimney sweeps to celebrate the occasion by covering themselves in foliage and flowers as a character called Jack of the Green or Jack in the Green, to coax coins from the crowd. The practice began to die out after 1875, when the Chimney Sweepers Act banned children from doing the work.

In America in the 1920s, 'Prohibition green' was the colour painted on the doors of speakeasies to let the cognoscenti know that liquor was available. This custom is one of the likelier explanations for the mysterious song 'Green Door,' a chart-topper for co-writer Jim Lowe in the US and Shakin' Stevens in the UK, in which the sleepless narrator is stuck out in the cold while people behind a green door play an old piano and laugh a lot.

At the end of the 1960s, legendary guitarist Peter Green, having come to loathe his fame and wealth, took to urging his bandmates in Fleetwood Mac to give away a lot of their money. When Mick Fleetwood and John McVie demurred – partly because, as the latter said, it was 'coming from a guy who'd just been given a lot of acid' – Green decided to quit. Before his departure he wrote 'The Green Manalishi (With the Two Prong Crown)', one of the most unusual singles ever to make the British charts. A top ten hit in 1970, the song was written after an LSD-induced dream in which he was dead and being barked at by a green dog that symbolised money and, by extension, the Devil.

As Michel Pastoureau has observed, 'In Egyptian painting the colour green is always seen as good and takes on various meanings: fertility, fecundity, youth, growth, regeneration, victory over disease and evil spirits.' The fertility god Osiris was portrayed with green skin, and Ptah, a deity associated with craft and creation, has a healthy green face on the tomb of the pharaoh Horemheb in the Valley of the Kings. A pale green eye shadow made from malachite doubled as a sunscreen for those Egyptians who had the status and/or the means to wear it.

Why is green the colour of American money? The US federal government did not print paper currency until the 1860s, when it needed to finance the Civil War. Using green ink on one side of the dollar bill was in part a measure to stop counterfeiting – cameras could not replicate the notes because, at that time, they could only take images in black and white.

When dollar bill design was standardised in 1929, the Bureau of Printing and Engraving stuck with green because, it says, 'the ink was plentiful, durable and conveyed stability'. Switching colours would have been problematic at this point because the

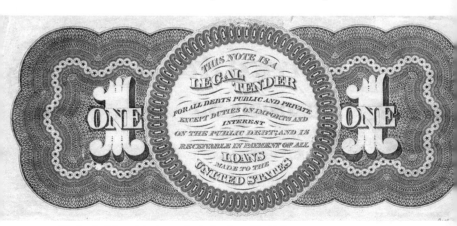

The back of the original greenback, first issued in 1862.

bills were universally known as 'greenbacks'. In recent years, to make the currency harder to fake, dollar bills have become more colourful, with subtle background shades of blue, a coppery orange and purple.

One day in 1786, the great Swedish-German chemist Carl Scheele (1742–86) was found lying dead at his workbench, in a cloud of toxic chemicals. He was 43 years old. His insistence on inhaling and tasting the substances he created had finally killed him. Renowned as the scientist who first identified oxygen, molybdenum, tungsten, barium, hydrogen and chlorine, he also gave his name to Scheele's green, a copper arsenite compound widely used to dye fabrics and wallpaper.

When Scheele invented the compound in 1775, he recognised that the arsenic content carried some risk. On the other hand, the colour could not be matched by any organic dyes. An arsenical rival to Scheele's green went on sale in 1814, in the German town of Schweinfurt, having been perfected by the chemists Russ and Sattler. Variously known as Schweinfurt green, Paris green (ostensibly because it was used to kill rats in the city's sewers), Vienna green and emerald green, the new pigment was more durable than Scheele's, and was favoured by artists such as Turner, Monet, Gauguin and Cézanne.

In Victorian Britain, Scheele's green became a menace – or was widely supposed to be a menace – thanks to the invention of a machine that created continuous lengths of wallpaper, and the repeal, in 1836, of the wallpaper tax. As Bill Bryson writes in *At Home*: 'By the late nineteenth century, 80 per cent of English wallpapers contained arsenic, often in very significant quantities. A particular enthusiast was the designer William Morris who was

Arsenic-laced wallpaper from William Morris – his 'daisy' design.

on the board of (and invested heavily in) a company in Devon that made arsenic-based pigments.'

Some people were alarmed. As early as 1839, the German chemist Leopold Gmelin called for Scheele's green to be banned. In 1862, arsenic-laced wallpaper was blamed, at an inquest, for the deaths of Ann Turner and three of her daughters in Limehouse, east London. By the 1870s, consumer concern had forced Morris to use arsenic-free greens. He did so with bad grace, complaining to a friend: 'As to the arsenic scare, a greater folly it is impossible to imagine: the doctors were bitten as people were bitten by the witch fever. My belief about it all is that doctors find their patients ailing, don't know what's the matter with them and in despair put it down to the wallpapers, when they probably ought to put it down to the water closet.'

It's possible that some people were made unwell by their green wallpaper, but the lethality of Scheele's green is doubtful. In 2005, research by William Cullen and Ronald Bentley concluded that the problematic wallpapers did not generate as much arsenic gas as had been previously thought, and that mould might have been responsible for the illnesses that had previously been blamed on toxic decor.

When Napoleon Bonaparte died on the remote island of St Helena, on 5 May 1821, aged 51 – some were quick to suspect foul play. The British had tried to assassinate him in 1800 and 1804, and Napoleon was convinced they had not given up, writing three weeks before his death: 'I die before my time, murdered by the English oligarchy and its assassin.'

There was no evidence of anything untoward, however, until the 1960s, when tests revealed arsenic in a lock of Napoleon's hair. Swedish dentist and amateur toxicologist Sten Forshufvud proposed that the Charles-Tristan Comte de Montholon, who had managed the emperor's household on St Helena, had poisoned the ex-emperor. The motive could have been jealousy (Napoleon had had an affair with the Comte's wife) or plain greed – the Comte stood to inherit two million Francs, as the largest single beneficiary of Bonaparte's will. (The practice of poisoning the wealthy was so widespread in the nineteenth century that arsenic was known as 'inheritance powder'.)

Another theory was put forward: the arsenic might have come from the green, gold and white wallpaper in Napoleon's bedroom at Longwood House. St Helena's humid, tropical air could have created a mould which, reacting with the paper, gave off a toxic gas containing an arsenic compound called trimethylarsine. Some members of his entourage did complain of 'bad air' in Longwood House. Yet, in 1982, researchers David Jones and Kenneth

Ledingham concluded: 'X-ray fluorescence measurements on this paper reveal enough arsenic to be capable of causing illness but probably not death.' It is generally thought that he died of stomach cancer.

The challenge of achieving a bright but consistent green bedevilled artists for centuries. In the 1670s, Dutch painter Samuel van Hoogstraten complained: 'I wish that we had a green pigment as good as a red or yellow. Green earth is too weak, Spanish green is too crude and ashes not sufficiently durable.'

Some relied on verdigris, a green compound that forms on copper as it weathers. But verdigris-based pigment often darkened with age, as Leonardo da Vinci cautioned in his notebooks: 'Verdigris with aloes, or gall, or turmeric makes a fine green and so it does with saffron or burnt orpiment, but I doubt whether in a short time they will not turn black.'

Green became less of a challenge in 1838, when Parisian colour man Pannetier discovered a formula to turn chromium oxide into *vert émeraude*, a vivid green with a bluish tint, known as viridian in English. The pigment became more popular after Orientalist painter Jean-Adrien Guignet perfected and patented his own recipe for viridian in 1859. Chrome green, a variable mix of Prussian blue and chrome yellow, was popular for a while – primarily because it was cheap – but artists found it unreliable. Viridian was adopted enthusiastically by the Impressionists and became one of Cezanne's favourite colours.

Cobalt green, a pigment devised by Swedish chemist Sven Rinman in 1780 by mixing cobalt oxide and zinc oxide, has never been

Manet's *The Balcony* (1868) makes extensive use of the new chrome green pigment – including a pure viridian parasol.

popular with painters, chiefly because it's expensive and mixes poorly with white. But researchers at the University of Washington have discovered that the pigment may prove useful in quantum computing. One of the basic applications of quantum computing is called spintronics – in a nutshell, by exploiting the spin (as well as the charge) of an electron it's possible to create computers that store and transfer data at previously unheard-of speeds. Spintronics was developed in the 1980s, but progress has been hampered by the fact that these devices could only work at −200 °C. Rinman's cobalt green, tests suggest, has particular magnetic properties that could enable quantum computers to function at room temperature – so the unloved pigment may play a part in ushering in another technological revolution.

The green flash – or green ray, as it's also known – is a rare optical event caused by the way the Earth's atmosphere bends and scatters light from the departing sun. In the right conditions

– defined in *National Geographic* magazine by Nadia Drake as a 'clear, unpolluted horizon free of clouds and haze' after sunset or before sunrise – only green wavelengths of light reach our eyeballs, as the rest are filtered out. When Drake glimpsed this phenomenon on the Hawaiian island of Kauai in January 2016, she noted: '"Flash" isn't quite the world I'd use to describe this phenomenon. The green colour was fleeting, to be sure, but it simmered rather than burst and oozed rather than erupted. It was more of a "green glow" or "green smear".'

According to an old Scottish tradition, Drake – having seen the green flash, ray, glow or smear – will never be deceived in love. This legend inspired Jules Verne's novel *Le Rayon Vert* (*The Green Ray*), which in turn inspired an Éric Rohmer film of the same name. In the novel, the heroine tells her kindly guardian uncle that she will not marry until she has seen the green ray, because it will then be 'impossible to be deceived in matters of sentiment'. When the flash does appear, she is too busy gazing into the eyes of her new love to notice.

The green ray finds a distant echo in Manx mythology. In an article in *The Times* on 10 September 1929, Mona Douglas wrote that the green flash was called *soilshey-bio* ('living light') in Manx, the ancient language of the Isle of Man. 'In several fragments taken down by me from Manx fisher folk, the "flash" was seen at sunrise on the morning preceding the wreck of one or more boats, sometimes by a relative of the men actually lost and in other cases by the men themselves who took the warning and withdrew from their fated enterprise.'

It is often suggested that a green flash immediately preceded Krakatoa's volcanic eruption in August 1883, which unleashed a force 10,000 times as powerful as the Hiroshima bomb, killing 36,000 people. Be that as it may, it's a fact that the layer of ash particles produced by Krakatoa did something far weirder – it turned the sky green.

On 27 August 1883, a *Manchester Guardian* reader identified only as 'J.T.G.' was on the deck of a steamer between Java and

Sumatra when, as he wrote, 'a squall was seen coming up from the south-west which as it came nearer we observed to be of a vivid green colour … The sea was also very green, about the colour of a well-kept lawn … The sky was continually of a green colour for some days and, steering west the sunrises and sunsets were indescribably beautiful, tinged with every shade of green not simply at the place of rising and setting, but thrown back on the rolling clouds all around the horizon.' The skies stayed green until the steamer reached the Red Sea.

In the entire history of Western art, has any garment caused more confusion than the luminous green dress worn by the apparently pregnant young woman in the picture traditionally known as *The Arnolfini Wedding*? Painted in 1434 by Jan van Eyck, this double portrait is an enigmatic creation, and scholars continue to debate various aspects of its imagery. It's now generally accepted, however, that the man in the picture is Giovanni di Nicolao Arnolfini, a Bruges-based Italian cloth merchant, and that the woman is Costanza Trenta, whom he married in 1426. Costanza died in 1433, however, which means that van Eyck's masterpiece is not so much a celebration of a wedding as a memorial to a deceased spouse.

And, contrary to appearances, the young woman is not pregnant – she's just the wife of a very rich man. Her ample fur-lined dress is an example of early fifteenth-century Netherlandish haute couture – the deeply pleated fabric is a deluxe wool that would have been as costly as silk, and it would have required a double-dyeing process to achieve that deep emerald tone. No expense was spared with the materials of the painting, either. The lustrous pigment of the dress is verdigris, which van Eyck – one of the pioneers of oil painting – mixed with linseed oil, a

The Arnolfini Wedding – painted by van Eyck in 1434 and still perfectly emerald.

technique that deepened the intensity of the colour. However, verdigris becomes transparent in oil, so van Eyck added a dash of pine resin to make the colour opaque. This in turn made the pigment prone to discolouration, a problem which the painstaking artist resolved by applying the paint in glazed layers, to keep it stable. The end result is an image that looks almost as fresh today as it did six hundred years ago.

If you are going to call yourself 'Special Forces' you need a special hat to reinforce your point. That was certainly how the United States Army Special Forces saw things when they were founded in 1952. Having searched through all the headgear they had accumulated, they opted for a rifle green beret for a retirement parade at Fort Bragg, North Carolina, in 1955. Unfortunately, after the US Air Force contingent was mistaken for a foreign delegation from a NATO ally, the top brass banned the berets. Special Forces members wore them surreptitiously until October 1961, when President Kennedy officially sanctioned their use by the USAF. As *Stars and Stripes* correspondent Forrest Lindley recalled: 'People were sneaking around wearing [them] when conventional forces weren't in the area and it was a sort of cat and mouse game. When Kennedy authorised the Green Beret, everybody had to scramble around to find berets that were really green. Some were brought in from Canada. Some were handmade, with the dye coming out in the rain.'

In 1995, when the mayor of Istanbul, Recep Tayyip Erdoğan, now the increasingly authoritarian president of Turkey, began painting the city's pavements green, it caused much consternation. As novelist Kaya Gens recalled in *Guernica* magazine in 2013: 'I wasn't surprised to see widespread panic in the city's westernised neighbourhoods. "The Sharia is coming!" people screamed on the streets as if a Martian or Godzilla himself had materialised nearby and were approaching with their green steps, sloshing their disgusting colour onto whoever they came across.' To a self-described 'cardigan-wearing citizen of a secular republic' such as Gens, green represented Islam and Sharia law. Erdoğan finally bowed to public pressure, repainting the pavements a cheery yellow.

The identification between green and Islam begins with the prophet Muhammad (571–632) and the Qu'ran (in which green is associated with vegetation, the spring and paradise). In the deserts where Muhammad began preaching, green would have been a powerful symbol of life. That may be why, according to some of

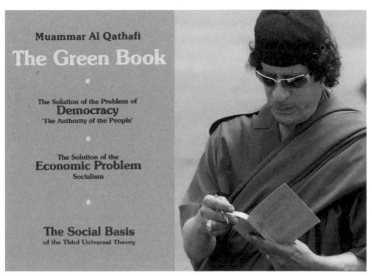

Gaddafi engrossed in his *Green Book*.

his companions, the prophet offset his white clothes with a green turban and green cloak. After Muhammad's death, green became the family's dynastic colour. In 1171, Saladin destroyed the caliphate founded by the Fatimids, who claimed descent from Muhammad's daughter, Fatima, and his cousin Ali, and adopted green for his Saracen armies.

Islamic green features, to varying degrees, on the flags of Afghanistan, Algeria, Azerbaijan, Comoros, Iran, Iraq, Jordan, Kuwait, Lebanon, Libya, Mauritania, Oman, Pakistan, Saudi Arabia, Sri Lanka, Sudan, Syria and the United Arab Emirates. In 1977, Muammar Gaddafi gave Libya an all-green 'Great Socialist' flag, a proclamation of the power of Islam which also, handily enough, promoted his *Green Book*, a collection of aphorisms published two years earlier.

PINK IS FOR BOYS

The use of 'pink' as a colour term may have arisen from the *Dianthus* group of plants, particularly the fragrant *Dianthus plumarius* (or garden pink). The petals of these flowers were said to be 'pinked' – a word meaning 'cut', 'notched' or 'pierced'. Pink as a signifier of a pale red colour can be traced back only to 1733 – when, oddly, the word also designated 'A yellowish lake pigment made by combining a vegetable colouring matter with a white base,' to quote the *Oxford English Dictionary*. And in *The Artist's Handbook*, Ralph Mayer describes a pigment called Dutch pink, which he defines as 'A fugitive yellow lake made from buckthorn (Avignon or Persian) berries'. How the word pink ever came to mean 'yellow' remains a mystery.

❖ ❖ ❖

Dianthus comes from the Greek for 'flower of God' and, according to Christian legend, the first carnations (*Dianthus*

Elizabeth I with a pink carnation in *The Hampden Portrait*, c.1560.

caryophyllus) sprang up on the spot where the Virgin Mary's tears fell as she witnessed Jesus's suffering at Calvary. In the *Madonna of the Pinks* (1507), Raphael portrays a young Mary playing with the Christ child, who holds carnations that serve as a harbinger of the Passion.

The carnation's association with the Virgin Mary may explain why Elizabeth I, who carefully cultivated her image as the Virgin Queen, is shown holding a carnation in the famous *Hampden Portrait*, dated to around 1560. The symbolic association with Christ's mother underlined the queen's role as head of the Church of England. Furthermore, the carnation was an element in what historian Roy Strong described as 'an enormously diffuse horticultural image in which the Queen, the kingdom, the spring, the garden and the flowers become inextricably intertwined'.

In Elizabethan England, 'pink' was sometimes used as a synonym for 'flower', in the sense of 'the best' (as in 'the flower of England'). This is how Mercutio uses it in Shakespeare's *Romeo and Juliet* (1597), telling Romeo: 'Nay I am the very pink of courtesy.' Romeo's reply, 'Pink for flower', uses the colour in a very different sense, alluding to women's genitalia.

In 2018, scientists from the Australian National University discovered a bit of bright pink pigment in 1.1-billion-year-old rocks that had been extracted from the Taoudeni Basin, in Mauritania, West Africa, ten years earlier. After the rocks had been pulverised in a lab, in an attempt to extract molecules from any ancient organisms trapped inside, graduate student Nur Gueneli mixed the powdered rock with an organic solvent, which promptly turned pink. As her colleague Jochen Brocks told the *Guardian*: 'I heard this screaming in the lab. [Gueneli] came running into my office and said "look at this" and she had this bright pink stuff.' The pigment, which comes from the chlorophyll of fossilised cyanobacteria, is at least 600 million years older than any previously known specimen.

One of the odder items in the Victoria and Albert Museum's collection is a knitted pink hat, known as a 'pussy hat', one of many worn by hundreds of thousands of protesters in the

Pink pussy hats at the Women's March on Washington, DC, 2017.

Women's March on Washington, DC, on 21 January 2017. The hats were a reference to Donald Trump's infamous assertion that women would let you do anything to them if you were famous, and that you could 'grab them by the pussy'. *Washington Post* journalist Petula Dvorak, who believed the hats trivialised the issues at stake, wrote: 'Please sisters, back away from the pink.'

In 1991, the Susan G. Komen Breast Cancer Foundation gave pink ribbons to the runners in its survivor race in New York. Within a year, the pink ribbon had become the official symbol of Breast Cancer Awareness Month after being featured on the cover of Condé Nast women's magazine *Self* and promoted by Estée Lauder. The pink ribbon is now the ubiquitous symbol of breast cancer awareness and research.

According to Edith Piaf's biographer Carolyn Burke, the singer wrote the lyrics of 'La Vie en Rose' ('Life in Pink') on a paper tablecloth at a café on the Champs-Élysées, for her friend Marianne Michel, who was complaining that she had nothing new to sing. At first, Piaf scribbled the lyrics 'Quand il me prend dans ses bras/Qu'il me parle tout bas/Je vois les choses en rose.' Michel suggested that the phrase 'les choses en rose' – 'the pink things' – would work better as 'vie en rose'.

'La Vie en Rose' became Piaf's theme song, selling a million copies in the US alone in 1947. Yet her actual life reaffirms the French saying 'Tout n'est pas rose' ('It's not all rosy'). Abandoned at birth by her mother and raised in a brothel, she was accused – but not convicted – of collaboration during the Nazi occupation of France, suffered three serious  car crashes, which exacerbated her dependence on alcohol and morphine, and died of an aneurysm caused by liver failure at the age of 47. The love of her life – the great French boxer Marcel Cerdan – died in a plane crash in 1949, aged 33, a little more than a year after their affair began.

In the summer of 1952, one muggy Memphis afternoon, Bernard Lansky spotted a shy, acne-scarred white teenager

Elvis Presley poses for a portrait in 1956 in Memphis, Tennessee.

peering through the window of his clothing store. As he recalled, 'It was very seldom I saw a white dude come down on Beale Street to look and see what was happening. He was interested in seeing what we had in the window.' Lansky's had prospered by selling yellow suits, pink sports coats, silk shirts and white shoes to Beale Street's pimps, gamblers and entertainers. Intrigued by the window shopper, Lansky invited him in and showed him round. 'I like all this, it's fantastic, when I get any money I'll buy you out,' said the 17-year-old, who was then working part-time as an usher at a nearby cinema. To which Lansky replied: 'Don't buy me out, just buy from me.'

The cinema usher was Elvis Presley and, as soon as he found fame and fortune, he took Lansky's advice. A year later, for his high school graduation, Lansky: 'put a pink coat on him, black pair of pants and a pink and black cummerbund.' Pink and black was, for a while, Presley's favourite colour combo. He wore a pink and black bowling shirt on *The Milton Berle Show* in 1956, when his hip-twitching rendition of 'Hound Dog' provoked such outrage that one Catholic newspaper headlined its review: 'Beware of Elvis Presley'.

In staid 1950s America, Presley's taste for pink shirts, pink jackets and pink Cadillacs was a deeply subversive act. There was nothing subtle about his preferred pink – it was akin to the 'hot pink' or 'shocking pink' invented by designer Elsa Schiaparelli. At that time and place, pink was a colour for women and blacks. Even the singer's relatives were shocked. As his cousin Billy Smith put it: 'Most of the family thought: "Well, why doesn't he just go down there and live with them?"' The flashiest symbol of his contempt for social norms was the pink Cadillac he bought his mother, Gladys.

Actress Jayne Mansfield adored pink cars, and also had a pink shag-pile carpet in her pink mansion and dyed her pets' fur pink. She did this because, she said, 'men want a girl to be pink, helpless and do a lot of deep breathing'. Barbara Cartland, the eccentric doyenne of British romantic fiction, wore coral pink gowns because, as she once told an interviewer, 'Who can be happy and pretty in grey?' Elaborating on her philosophy of dressing, she insisted that 'no Englishwoman should wear beige or brown, because it makes them look like a baked potato'.

In Marcel Proust's *Du côté de chez Swann*, the first volume of the monumental *À la recherche du temps perdu*, the narrator, as a boy, meets an enchanting 'lady in pink', who will turn out to be Odette de Crécy, the courtesan who marries Charles Swann. The sensuality of Madame Swann is later encapsulated by the sleeves

Swann's Way: Tiepolo's signature pink on his Venus and Vulcan.

of her gown, which are of a colour 'so peculiarly Venetian as to be called Tiepolo pink'. A pale and airy pink was a signature colour of the Venetian artist Giovanni Battista Tiepolo (1696–1770), of whom Roberto Calasso wrote, in his book *Tiepolo Pink*, 'Every fibre of his painting is erotic.' This is no overstatement – even Tiepolo's angels look nubile, and his goddesses flaunt themselves on clouds that resemble vast pink beds.

It is said that the young Madame de Pompadour, determined to make an impression on King Louis XV, attracted his attention by twice driving past him when he was hunting in the Forests of Sénart – once riding in a pink phaeton, while wearing a blue dress; then in a blue phaeton, dressed in pink. Such was her love of pink that the Sèvres porcelain factory created a new tone, Rose Pompadour, in her honour.

Mamie Eisenhower wore a pink gown, embroidered with more than 2,000 rhinestones, to her husband's inaugural ball in 1953. She was so fond of the colour that, during Ike's presidency, the White House – with its pink kitchen and pink furnishings – became known as the 'Pink Palace'. Her penchant for pink was said to signify that she would, in the words of *Time* magazine, 'lend a new warmth to the affairs of the presidency'. The shade known as 'Mamie's pink' was for a while one of the most popular decorating colours in the USA.

The shocking pink of the dress worn by Marilyn Monroe as she sang 'Diamonds Are a Girl's Best Friend' in *Gentlemen Prefer*

Marilyn Monroe introduces the world to shocking pink (best worn with diamonds...).

Blondes owed its creation to an actual diamond – a stone called the Tête de Belier ('Ram's Head'), once owned by Daisy Fellowes, heiress to the Singer sewing machine fortune. One of Fellowes' favourite fashion designers, Elsa Schiaparelli, was so taken with the diamond's sparkling colour she concocted a super-saturated, glaring shade of magenta, which she called shocking pink. First featured in her 1937 collection, the colour became synonymous with the Schiaparelli label. On 11 June 2010, Monroe's pink dress was sold at auction for $370,000.

A precious gem also inspired one of the pinkest cultural icons of the 1960s and 1970s. The *Pink Panther* movie (1963) took its title from a flawed diamond which, when held up the light in a certain way, showed the figure of a springing panther. In 1969, NBC launched a mildly surreal, largely silent, comic cartoon series centred on a pink big cat. Animator Friz Freleng, who also developed Bugs Bunny, Porky Pig and Sylvester the Cat, regarded the Pink Panther as his finest creation.

Three years after Schiaparelli invented shocking pink, the lavender-mauve-grey Mountbatten pink was created, at the initiative of Louis Mountbatten, then Britain's First Lord of the Admiralty, who believed it would camouflage ships more effectively than grey. It did not.

The first half of the twentieth century was a particularly fertile period for new shades of pink. Varieties of pink that date from this period include: Amaranth pink (first described in English in 1905), the light Baby pink (1928), orangeish Congo pink (1912), the medium light rosé Cameo pink (1912), the moderately red Tango pink (1925), the bright purplish Persian pink (1923), the darker purplish China pink (1948), shocking pink (created in 1937) and the grayish Silver pink (1948). One of the past century's most ubiquitous pinks was Pantone 219 C, the colour used by Mattel for the Barbie doll.

Why is the Argentinian president's executive mansion painted pink? Legend has it that during the presidency of Domingo Sarmiento (1868–74) the façade of the Casa Rosada – 'Pink House' – was painted with a mixture of cows' blood and paint, to resist the humidity for which Buenos Aires is notorious. The more likely explanation is that the colour was chosen to defuse political tension by blending the Federalists' red with the white of the Unitarians, the party to which Sarmiento belonged.

The idea that pink is a girl's colour is not as deeply ingrained as one might think. In Louisa May Alcott's *Good Wives* (1869), Amy is advised to put 'a blue ribbon on the boy, and a pink ribbon on the girl, French fashion', and yet, in 1918, the august American trade publication Earnshaw's advised its readers: 'Pink, being a more decided and stronger colour, is more suitable for

Pink is for cowboys in this Sears Christmas catalogue from 1955

the boy, while blue, which is more delicate and dainty, is prettier for the girl.' In the following decade, leading American department stores Filene's, Best & Co, Halle's and Marshall Field's all recommended that boys be dressed in pink.

One of popular culture's most enduring heroines, *Alice in Wonderland* has been portrayed in a blue dress since Macmillan's deluxe edition of the book in 1911. It wasn't until the 1940s that America's department stores and manufacturers decided that customers preferred girls to wear pink. Yet in the words of Marilyn Read, a professor at Oregon State University who researches the design of children's environments, 'Pink is a colour that boys like until they're told not to.' As Read points out, 'no significant research . . . demonstrates a difference in colour perception between boys and girls', and several studies have suggested that boys' and girls' first colour preferences are for blue or purple.

In 2016, a government drive to make Chinese cities more 'female friendly', by creating wider parking spaces for women with pink borders and a pink skirted figure in the middle, sparked accusations of sexism on social media. A service area manager in the city of Hangzhou defended the scheme, saying, 'women drivers' skills are not superb'.

'In the eighteenth century, it was perfectly masculine for a man to wear a pink silk suit with floral embroidery', says fashion scholar Valerie Steele. By the time F. Scott Fitzgerald wrote *The Great Gatsby*, attitudes had changed – Jay Gatsby's pink suit marks him out as a flashy arriviste. Informed that Gatsby is 'an Oxford man', his upper-class nemesis Tom Buchanan replies scornfully: 'Like hell he is! He wears a pink suit.' Nick Carraway's final glimpse of his friend, in a 'gorgeous pink rag of a suit', is a poignant reminder that Gatsby's attempt to refashion himself as a member of the American aristocracy has failed.

Pink is one of the predominant colours of the interstellar clouds of dust and ionised gases known as nebulae. Astrophysicist Frank Summers explains: 'The most abundant element in the universe is hydrogen and, when heated by hot stars to several thousand degrees, it glows with a soft red light that appears pink. Other nebulae appear blue because short wavelengths of blue light are more easily reflected than longer wavelengths (like red light) so nebulae that reflect the light of nearby stars will appear blue.'

Summers adds: 'My favourite cosmic colour is the pink of Hydrogen Alpha, emanating from star-forming regions. This gas, glowing at several thousands of degrees, is the nebular cocoon around a batch of newly born stars. The pink glow of a stellar nursery is a warm colour that, like the green of springtime, indicates the continual progression of stellar life.'

Time magazine coined the word 'pinko' in 1925 and, around that time, the *Wall Street Journal* criticised the 'parlour pink' followers of progressive senator Robert La Follette. Spinning off red's association with Communism, 'pinko' also implied that La Follette and his ilk were effete and unmanly.

As early as 1934 the Gestapo began compiling 'pink lists' of known homosexuals. One of the most prominent homosexuals

Memorial plaque to the homosexual prisoners in Buchenwald concentration camp, near Weimar, Germany.

in Germany was Ernst Röhm, who, as head of the SA storm-troopers, became a political liability for his erstwhile friend and ally Adolf Hitler. The Night of the Long Knives, on 30 June 1934, in which Röhm and his cohorts were executed on Hitler's orders, was portrayed by the Nazi regime as a triumph over moral turpitude. Homosexual inmates in concentration camps were forced to wear a triangular pink triangle. An inverted pink triangle has since become a symbol of gay pride.

The blossom of *Prunus serrulata*, the Japanese cherry – known in Japan as sakura – is the country's unofficial national flower, and most of the 600 varieties are pink (some are white, others almost crimson). The first mention of sakura in Japanese literature occurs in the *Nihon Shoki*, completed in 720 CE, in which a petal falls into the *saké* cup of Emperor Richū. In literature and songs, the cherry blossom is revered as symbol of beauty and renewal, but also – because they bloom for only a few weeks – of the transience of life. A popular couplet by an anonymous Japanese poet runs: 'It is because they scatter without trace that cherry blossom delights us so/For in this world lingering means ugliness.'

Some scholars say that a fallen cherry blossom came to symbolise young samurai who had fallen in the service of their master. Others argue that this association was exaggerated retrospectively by later governments, particularly the one that led Japan into the Second World War. Yet there is age-old evidence of the linkage – for example, the medieval proverb 'the [best] blossom is the cherry blossom, the [best] man is the warrior'. Some samurai committed *seppuku*, ritual suicide, under the cherry blossoms. In so doing, they were deemed to have died at a moment of superlative beauty – an ideal death.

Servants of the Japanese state such as scholar, poet and mandarin Matsudaira Sadanobu prized sakura for another reason: the cherry blossom was exclusively Japanese. As Sadanobu wrote in 1818: 'Even knowing that the cherry blossom was supposedly unique to our country, I thought it must surely exist in China and did a good deal of searching, only to find no Chinese paintings of cherry blossoms and no Chinese poems that referred to them.' Sadanobu's nationalist interpretation of sakura influenced

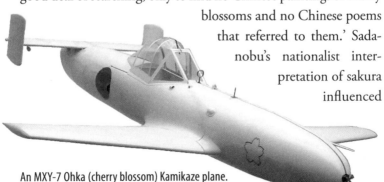

An MXY-7 Ohka (cherry blossom) Kamikaze plane.

subsequent state propaganda promoting the country's *kokutai* or 'national essence', such as the 'Song of Young Japan', an anthem which celebrated warriors who were 'ready like the myriad cherry blossoms to scatter'.

In 1930, the Sakurakai (Cherry Blossom Society), an ultra-nationalist group within the Japanese Army, plotted to install a totalitarian regime under Emperor Hirohito. After two failed coups d'états the society was dissolved in 1931, yet it remained influential: during the Second World War, the Sakurakai leader, Lieutenant Colonel Kingoro Hashimoto, was in charge of indoc-trinating the young.

As Ian Buruma writes in *A Japanese Mirror*, the cult of cherry blossoms was, in part, 'a cult of death'. At its most extreme, that cult was expressed by the sakura that many kamikaze pilots wore as sprigs on their uniforms or painted on the side of their planes before their first and last missions. (*Kamikaze* means 'divine wind', referring to the typhoon that fortuitously dispersed a Mongol fleet in 1281.) The letters and poems those doomed pilots left behind, Buruma notes, often refer to sakura. Before his mission, one 22-year-old pilot wrote:

> *'If only we might fall*
> *Like cherry blossoms in the spring*
> *So pure and radiant.'*

To remain pure and radiant meant climbing into a plane that was loaded with 2,600 pounds of dynamite in the nose, strapping yourself in with no eject button, and slamming the plane, at a speed of 600mph, into an American warship. From their first attack, at the Battle of Leyte Gulf in October 1944, until the end of the war, the kamikaze sank at least 34 ships (some sources say

47) and damaged many more. At the Battle of Okinawa in 1945, 4,800 Americans were killed in kamikaze attacks, almost twice as many as died at Pearl Harbor.

Yet the symbolism of cherry blossoms in Japan is too complex to be monopolised by any single cause or faction. As the poet Motoori Norinaga (1730–1801) wrote: 'If I were asked to explain the Japanese spirit, I would say it is wild cherry blossoms growing in the morning sun.' The business of tracking the blossoms preoccupies much of Japan from March to May every year, as families and workmates gather at *hanami* (blossom-viewing) parties, celebrating the renewal that the sakura embodies. A cherry tree provided a poignant postscript to the Great East Japan Earthquake and tsunami in 2011. The survival of a 900-year-old specimen, thirty miles from the Fukushima nuclear power plant, transformed a symbol of fragility into an emblem of resilience.

Have you ever seen pink elephants when inebriated? Such hallucinations feature in a bizarre sequence in Walt Disney's animated classic *Dumbo* (1941). After inadvertently drinking water spiked with champagne, the flappy-eared pachyderm dreams of pink elephants singing, dancing and forming a marching band. Waking up in a tree, Dumbo discovers he can fly. 'Seeing pink elephants' was a popular euphemism for getting so drunk that you had visions. In 1896, the Western author Henry Wallace Phillips referred to a drunken man seeing a 'pink and green elephant and a feathered hippopotamus'. Other writers preferred blue monkeys or yellow and green giraffes, but Jack London, in his autobiographical novel *John Barleycorn* (1913),

Hanami – celebrating the cherry blossom in Kyoto's Maruyama Park.

portrayed the type of drinker 'who sees, in the extremity of his ecstasy, blue mice and pink elephants'. Today, a pink elephant is the logo of the Belgian beer Delirium Tremens, and is the name of a vodka-based cocktail.

In 1979, at a clinical ecology seminar, Alexander G. Schauss proposed that a rather lurid bright pink, dubbed Baker-Miller pink after the naval officers who had tested it in their cells, made prisoners less aggressive. The evidence seemed compelling. After the cells at the Seattle Naval Correction Center

A pretty idea? Non-violent pink ambience trialled at Seattle Naval Correction Center.

were painted a particular shade of pink, there wasn't a violent incident for 156 days. Schauss explained: 'Even if a person tries to be angry or aggressive in the face of pink, he can't. The heart muscles can't race fast enough. It's a tranquilising colour that saps your energy.' Inspired by such results, several nurseries, drunk tanks and away-team dressing rooms at college stadiums were painted pink. Unfortunately, subsequent studies have not supported Schauss's theory.

Even so, the legend of Baker-Miller pink endures. In January 2017, model Kendall Jenner, half-sister of Kim Kardashian, revealed that she had painted her living room pink as part of her weight loss regime. 'Baker-Miller pink is the only colour scientifically proven to calm you and suppress your appetite. I was like: I need this colour in my house!' she trilled.

Pink is indelibly associated with arguably the world's most authoritative sports newspaper, *La Gazzetta dello Sport*. Founded in 1896, it was printed on green paper, then yellow and white, before finally – and irrevocably – turning pink in 1899. The choice proved immensely significant when its owner Emilio Costamagna launched the Giro d'Italia on 13 May 1909. The association between the paper and the cycle race was strengthened in 1931, when the organisers began awarding a pink jersey to the race leader. Today's circulation is just under 400,000, less than half of what it was thirty years ago, but there's barely a café in Italy that doesn't have a copy for its customers to peruse, making *La Gazzetta* the country's most-read newspaper.

❖ ❖ ❖

Juventus, Italy's most popular football team, wore pink shirts from 1897 to 1903. There are various tales about how they came to play in black and white, but it seems that Juve's star player at that time, an Englishman called Tom Savage, was unhappy about playing in pink, and asked a factory back home if they could supply a replacement. The factory dispatched a batch of shirts with the black and white stripes of Savage's favourite team, Notts County. The *Bianconeri* – 'black and whites' – still have a pink change strip. In 1907, another Italian club, Palermo, chose pink and black shirts because, according to founding member Count Giuseppe Airoldi, the colours would

represent the sadness of defeat and the sweetness of success – a suitable combination for a team that had experienced so many ups and downs.

Convinced that gender-specific marketing was lowering the aspirations of young girls, twin sisters Emma and Abi Moore launched the Pinkstinks campaign in Britain in 2008. Baroness Morgan of Drefelin, then a junior children's minister, supported them, saying: 'It's extremely important that girls have a chance to play with trucks and trains and wear blue … we shouldn't be defining how people are looked after by the colour of their toys.'

The debate over what some critics call 'colour apartheid' has raged for decades. In the 1970s, for many Women's Liberationists – as the media invariably described them back then – pink represented an infantile, Barbie-fied version of femininity that existed primarily to please men. At the same time, many American women rallied behind the term 'pink collar' as they campaigned to improve their lot in offices and factories.

In 1982, a group of American women founded CODEPINK, a grassroots movement for peace and social justice. The group's name referenced the Department of Homeland Security, which did not feature the colour at all in its threat alert system.

In the Indian state of Uttar Pradesh, pink has been adopted by an all-female vigilante movement known as the Gulabi Gang (*gulabi* being Hindi for 'pink'), which was founded in 2006 by Sampat Pal Deva, to protect women against domestic violence. The group is controversial – it prefers to stay within the law, but members

The Gulabi Gang, led by Sampat Pal Devi, rally against violence against women in the Uttar Pradesh city of Banda, 2009.

have been known to beat rapists with large bamboo sticks – but, long before the advent of #MeToo, it attracted thousands of women and challenged India's patriarchal culture. Gang members wear pink saris so they can easily identify each other at rallies.

As a teenager in Pennsylvania in the mid-1990s, Alecia Beth Moore changed her stage name to P!nk, partly because many of her friends told her she reminded them of the Mr Pink character (played by Steve Buscemi) in Quentin Tarantino's *Reservoir Dogs*. Yet the singer-songwriter's choice was also a political statement: that feminists can wear pink. That message became more powerful as she was elevated by the media – and her commercial success – to 'pop royalty', influencing everyone from Christina Aguilera and Katy Perry to Lady Gaga.

In the 1960s, American racing driver Donna Mae Mims used pink to challenge motor racing's macho stereotypes. When she started racing in 1961, her husband Mike vetoed the idea of re-painting her Corvette in that colour, so Mims settled for wearing a pink helmet and overalls, having 'Think Pink' emblazoned on her cars, and branding herself as the 'Pink Lady'. In 1963, she raced in a pink Austin-Healey Bugeye Sprite and became the first woman to win a Sports Car Club of America national championship. As writer Lynn Peril put it, Mims' choice of colour signalled that she 'may have had the audacity to compete against men and win, but helped deflect criticism and reminded observers that she was, at heart, a girl like any other'.

On 6 October 2009, Mims died from a stroke at the age of 82. True to form, she requested that her body be placed in the driver's seat of a 1979 pink Chevrolet Corvette to receive visitors at her funeral.

In the United States, arms manufacturers such as Ruger, Glock, Mossberg and Remington have all produced pink guns to appeal to women and teenage girls. As the sales pitch for the Remington 870

Express Compact Pink Camo 20-gauge 21-inch shotgun puts it – 'You wanted some "wow" factor for young ladies, and we answered.' The girl-friendly firearm offers the 'perfect length, weight and balance for smaller-stature shooters'.

❖ ❖ ❖

When British artist Stuart Semple says, 'No one has ever seen a pinker pink' than his fluorescent pigment, he means it. Yet he would probably never have created the colour if he had not been incensed by the deal that gave Anish Kapoor exclusive rights to use Vantablack, the darkest shade of black ever devised, for artistic purposes.

Vantablack is not actually a paint or pigment at all – it is essentially a molecular trap for light. Developed by British company Surrey NanoSystems, it is made by using carbon nanotubes that are one ten-thousandth of the width of a human hair. These tubes are grown, as if in a carbon field of grass, and the light that falls between the tubes gets bounced around, absorbed and converted into heat. In other words, light gets in, but it can't get back out. It absorbs 99.965 per cent of light.

Creator Ben Jensen came up with a way to make Vantablack sprayable, so it could be used as a paint: BMW duly launched a Vantablack model in the autumn of 2019, billing it as the 'darkest car in the world', which surely cannot make our roads safer.

Like many artists, Semple was keen to paint with this new pure black. 'Black is the holy grail for artists,' he says, 'because it is no colour, and that is the starting point.' Likening Kapoor to the kid at school who wouldn't share his felt pens, Semple retaliated by crowdsourcing a new 'people's Vantablack' (not as dark as Surrey NanoSystems' colour but darker than most) and developing a vibrant pink that anyone – except Kapoor or his associates – could buy from his website.

Somehow, Kapoor did get his hands on the pinkest pink, dipping his finger in it and posting the image on Instagram. Ironically, Kapoor's 'Up yours!' sent sales soaring. 'All of a sudden we were getting thousands and thousands of orders. We ended up making pink paint for four weeks. It took over our lives,' says Semple. He has subsequently launched Black 3.0, which Kapoor and his associates are also not officially allowed to buy.

But why pink? Semple says that, in some way, it goes back to his first sight of Van Gogh's *Sunflowers* at the National Gallery when he was eight years old. 'I'd never seen colour that vivid,' he

Stuart Semple's patent pink: a message from Anish Kapoor.

says. 'I was literally awestruck. My mother found me standing by the painting, shaking.' It was a genetic rather than an artistic trait that led Semple to create his own pigments. Allergic to oil paints, he was forced to work with acrylics, which, he soon realised, severely restricted his palette. ('It stands to reason, really, oils have been with us for centuries, and acrylic paints only for decades'). Determined to give himself a wider choice, he began experimenting to see if he could mix acrylics that were as intense as the colours he imagined.

One of his first creations was the vibrant pink powder paint which reflects visible light so powerfully it looks fluorescent and, Semple says, 'goes bonkers under UV/black light'. He admits, 'We're not actually sure if this is the pinkest pink ever but it's as pink as we could make it and we've not seen anything pinker.'

BROWN
IN TOWN?

*'I cannot pretend to feel impartial about colours.
I rejoice for the brilliant ones and am genuinely
sorry for the poor browns.'*
Winston Churchill

What is the world's ugliest colour? In December 2012, the Australian government introduced plain cigarette packets, featuring pictures of rotting teeth and tumours on tongues, coloured in a drab dark brown known as Pantone 448 C. The hue was chosen by an advisory group of academics and market researchers, who had been given three months to design a packet to deter smokers. Research agency GfK Bluemoon conducted seven studies involving more than 1,000 regular smokers, aged 16 to 64, and decided that Pantone 448 C was the least appealing colour. Asked to describe what the shade made them think of, participants used such words as 'death', 'dirty' and 'tar'.

PANTONE®
448 C

Is Pantone 448 C the world's ugliest colour? Or would it make a nice mocha?

Lime green, white, beige, dark grey, dark brown, medium olive and mustard had all been tried, tested and rejected. The evidence was so compelling that the British and French governments have adopted the same colour in their campaigns.

Pantone was slightly put out by all the 'Is this the ugliest colour in the world?' headlines. As the ultimate authority on categorising the spectrum, the company officially insists that there are no 'ugly colours' and that 448 C is redolent of 'deep, rich earth tones'. But the Australian project would seem to support the 'ecological valence theory' of colour preference, expounded by Stephen E. Palmer and Karen B. Schloss in 2010. Their study, which asked a group of forty-two Americans to select their favourites from thirty-two chromatic colours, concluded that the participants' choices were generally determined by whether they liked the objects they associated with those colours. In plain terms, they liked blue (which reminded them of sunny skies) and green (which they connected with spring), but disliked brown because they associated it with rotten food, mud and excrement. On the other hand, another study has suggested that the word itself might be the problem. When people were shown two cards of exactly the same colour but named differently, many more liked 'mocha' than liked 'brown'.

The Australian government is not the first organisation to exploit Pantone 448 C's repellent qualities. In the 1960s, when an American department store chain asked consultant Faber Birren to discourage staff from taking long restroom breaks, he had the facilities painted a shade very similar to 448 C. It worked.

In fourteenth-century England, the lower classes were legally obliged to live in brown. In 1363, under King Edward III, a sumptuary law was passed to stop people spending lavishly on clothes and to make the difference between the social classes blatantly obvious. The paupers were told: 'Carters, Ploughmen, Drivers of the Plough, Oxherds, Cowherds, Shephards … and all other Keepers of Beasts, Threshers of Corn and all Manner of People of the Estate, and all other People, that have not Forty Shillings of Goods … shall not take nor wear no Manner of Cloth, but Blanket and Russet of Twelve-pence.'

The oldest surviving depiction of Saint Francis, from Subiaco Abbey, near Rome, c.1228.

Russet was a coarse woollen cloth which, when dyed with cheap, adulterated or very little madder, looked reddish-brown. As the shade of robe worn by the Franciscan order (which came to England in 1224), it also came to represent virtue and humility. Such associations resonated with Oliver Cromwell, who wrote to an associate in 1643: 'I had rather have a plain russet-coated captain that knows what he fights for, and loves what he knows than that which you call a gentleman and is nothing else.'

In the mid-1950s, as historian David Kynaston noted in *Family Britain*, homeowners began to repaint their brown interiors – useful, primarily, for hiding dirt – with Dulux white, as if simply changing the colour would usher in a brighter future. To be fair, though, in the 1970s brown was often used – particularly on wallpaper – in combination with brighter colours such as orange. After the psychedelic 1960s, the colour represented a return to a more grounded vibe. Car manufacturers promoted brown, too. One blogger complained, in retrospect: 'Seemingly everything that British Leyland offered was available in assorted shades of anaemic dog dirt.' The colour could occasionally look cool on cars – for example, on the brown Ford Consul GT driven by Detective Inspector Jack Regan in the iconic TV cop show *The Sweeney*.

The most ubiquitous man-made shade of brown is probably khaki, first worn in the 1840s by the British Army in India and later adopted by many other armies. In 1846, in Peshawar (now in Pakistan), Lieutenant-General Sir Harry Lumsden was looking for a suitable uniform for a new regiment. Aiming to make the men 'invisible in a land of dust' (*khaki* is the Urdu

The Indian Corps of Guards blending in, 1891.

word for 'soil-coloured'), he bought loads of white cotton from a bazaar in Lahore and ordered it to be soaked and rubbed with mud from a local river and cut into tunics and trousers. If there weren't enough mud and water nearby, coffee, tea, soil and curry powder would suffice. In 1902, a darker, greener shade of khaki became the British Army's standard field battle dress. The American Army later adapted the colour to olive green, but the original khaki lives on in chino trousers, a look initially adopted by preppy American students in the 1950s.

❖ ❖ ❖

Many shades of brown are named after animals, such as beaver, camel, chamoisee, coyote, fallow, faun, lion, seal brown, sepia (named after the cuttlefish from which the pigment is derived) and *taupe* (French for mole). Trees are an obvious source of browns, although probably not as many as you might expect: chestnut (and, from the French, maroon), mahogany, wenge and wood brown. The oldest source of brown is, of course, the earth, from which we get such pigments as burnt sienna (originally dug out of the ground in quarries near Siena and heated to darken it), Cassel earth, Cologne earth and umber.

Brown's most conspicuous corporate champion today is United Parcel Service, the global delivery service, which adopted it for its trucks and uniforms in 1916. Company founder James E Casey had wanted to use yellow, but one of his partners, Charlie Soderstrom, said they would be impossible to keep clean. Why not, he proposed, choose a shade of brown similar to the Pullman railway cars, a byword for 'class, elegance and professionalism', with their

The glamour of brown on the Pullman Golden Arrow and Orient Express.

luxurious and spacious sleeping berths? UPS adopted the shade and trademarked it – as Pullman brown – in 1998, almost thirty years after the collapse of the original railway car business.

In certain sections of British society, brown shoes are still not quite acceptable for men in the workplace. Back in the 1980s, a financial broker told the *Guardian* in 2016, 'If you were wearing brown shoes you'd get booed off the trading floor.' The dress code is not as strict nowadays, but one executive selling to the financial sector recently said: 'I'll put on my black shoes if I'm meeting a client. You're looked at as a bit of a spiv if not.' A recent memoir by Sasha Swire recalls prime minister David Cameron reproaching her husband Hugo, a Foreign Office minister, by looking at his shoes and asking: 'Brown in town?'

Brown shoes did, though, inspire Franz Zappa to write the song 'Brown Shoes Don't Make It', a commentary on a presidential fashion choice. As Hugh Sidey reported in *Time* magazine in 1967: 'When Lyndon B. Johnson once showed up for morning ceremonial duties in a grey suit and brown shoes, the people travelling with him were immediately alert for a change in the day's doings. Johnson was fastidious about the trappings of office, and even the slight dissonance of brown and grey hinted at a change of mood or schedule. Sure enough, LBJ jetted off secretly to

Vietnam.' During the flight, he changed into a rancher's outfit to match his footwear.

The brown shirts worn by Ernst Röhm's thuggish Sturmabteilung ('Storm Detachment') certainly helped bring the colour into disrepute. Their choice of uniform had arisen from an accidental discovery: the SA found a large stock of surplus shirts, originally ordered for Germany's colonial troops and left over after the First World War.

Brown envelopes can have a whiff of corruption about them. This was certainly the case with Neil Hamilton, a former Conservative minister, who in 1999 lost his libel action against Mohamed Al-Fayed, the owner Harrods, who had stated in a TV interview that Hamilton had demanded payment for asking question in Parliament on his behalf. Al-Fayed's former secretary, Alison Bozek, testified that he had put a 'wodge' of £50 notes into a white envelope, written the politician's name on it and given it to her to leave for collection in Harrods' Park Lane office. Because there was money in it, she put the white envelope in a brown envelope. Hamilton insisted: 'I never received a penny from Mr Fayed.' He lost the case.

In the 1960s and 1970s, many British football managers would collect illicit payments in brown envelopes at motorway service stations, by way of thanks for sanctioning transfer deals. The enduring appeal of a digitally untraceable brown envelope in football was confirmed in 2012, when an official in the Bahamas claimed he had been given envelopes stuffed with $40,000 in $100 bills to back Mohammed bin Hammam for FIFA president.

Investigators believed that up to twenty-five FIFA delegates had been offered the same deal.

Ugly, brown-skinned and hairy, the household spirits known as brownies in Scottish folklore have one redeeming virtue: they come out at night and do the household chores while the occupants are asleep.

A variant of these spirits is the titular protagonist in Scottish author James Hogg's *The Brownie of Bodsbeck* (1818). Initially presented as a deformed, ghostly monster, the Brownie is finally revealed to be John Brown, the battle-scarred, martyred leader of the Covenanters, a movement formed to protect the Presbyterian Church of Scotland against external interference, especially by the Stuart monarchs. Hogg plays a similar game in his short story

Palmer Cox's brownies, advertising their eponymous camera..

'The Brownies of Black Haggs' (1828), in which a mysterious creature called Merodach defies the evil lady of the house.

In North America, Palmer Cox's illustrated poems and cartoons about brownies spawned a lucrative merchandising business. The spirits' popularity is reputed, in 1900, to have prompted George Eastman to name his new, simple and cheap camera the 'Brownie'. It's a nice idea but it is surely no coincidence that it was invented by a man called Frank A. Brownell. What is not in dispute is that, in 1919, the youngest section of the Girl Guides movement in Britain changed its name from Rosebuds to Brownies, inspired by Ewing's story, in which two lazy boys are taught to become helpful 'brownies'. Given that rosebud has, over the years, acquired certain sexual connotations, it was probably a good call.

Umber, which is found in Neolithic cave paintings, is one of the oldest pigments known to humankind. Though some have argued that the name signifies that the pigment was originally made from earth in the Italian region of Umbria, it's likelier the word derives from *ombra*, Latin for 'shadow'.

The paintings of Michelangelo Merisi da Caravaggio made dramatic and innovative use of extreme tonal contrasts, placing brightly lit figures within a field of brown-saturated shadow. His chiaroscuro technique (from Italian *chiaro*, meaning 'light', and *oscuro*, 'dark'), had an extraordinary influence on his fellow painters, most notably on Rembrandt van Rijn, who learned about the Italian maestro's art from Dutch painters who had visited and worked in Italy. In Rembrandt's later works, brown became the dominant tone. As Philip Ball observes in *Bright Earth*: 'By the end of the 1650s, Rembrandt was using just a half-dozen or so pigments, mostly of dull earthly tones.' The pervasive gloom was not to everyone's taste. Rembrandt's near contemporary Gerard de Lairesse, for example, criticised him for 'seeking a mellowness, decayed into the ripe and rotten'. It's possible that financial considerations played a part in the restriction of Rembrandt's palette: he was left almost destitute after being declared bankrupt in 1656, and earth colours were probably cheaper than other pigments. But there is also an artistic truth to the darkening of Rembrandt's paintings. His later works are literally and psychologically darker,

Rembrandt's *Night Watch* – a classic example of his mastery of chiaroscuro.

and this surely reflects the pain of his final years, when he was impoverished and grieving the death of his son Titus (at the age of 27). When he died, just a year after his son, in 1669, his death went unreported in Amsterdam.

The Dutch artist Anthony van Dyck, court painter for Charles I, was so fond of – and adroit at using – a particular shade of brown that it became known as Vandyke brown after his death. According to Ralph Mayer, the colour was compounded of 'clay, iron oxide, decomposed vegetation (humus) and bitumen'. Depending on which authority you believe, it is either very similar to – or exactly the same as – Cassel earth and Cologne earth. In Belgium in the nineteenth century, the colour was known as Rubens brown. As a young man, van Dyck had studied in Antwerp under Rubens,  who described him as 'the best of my pupils'. Rubens liked to mix gold ochre into Cassel earth to warm it up. Van Dyck's brown is slightly drabber, but was perfectly suited to the shadowy landscapes that were then in vogue. The subdued colour spectrum was perpetuated by artists such as Reynolds and Gainsborough and, in an increasingly stodgy form, had become so dominant by the middle of the nineteenth century that the Impressionists mocked it as 'brown gravy'.

The use of 'brown study' to describe a melancholy, solitary reverie was part of the English and American vernacular in the nineteenth century, and can be found in the work of writers such as Charles Dickens, Louisa May Alcott and Sir Arthur Conan Doyle. Its first known appearance, however, was considerably earlier – in *Dice-Play*, a book published in 1532, the reader is advised that 'Lack of company will soon lead a man into a brown study.' The phrase 'brown study' as a code for feeling blue has since fallen out of fashion.

'I will pass through all thy flock to day, removing from thence all the speckled and spotted cattle, and all the brown cattle among the sheep, and the spotted and speckled among the goats: and of such shall be my hire.'
Genesis 30:32

This is not one of the most memorable lines in the King James Bible but, for linguists, it's a fascinating instance of the difficulties of determining exactly what was signified by the colour terms of ancient languages. The word brown appears four times in quick succession as Jacob and his employer Laban discuss how to divide up a herd. Published in 1611, the King James Bible was the first to translate the Hebrew word *khoom* as 'brown'. Aramaic and Latin versions translated it as 'dark'. Brown is not used in the Great Bible of 1539, the first authorised version in English.

One particular shade of brown almost stopped synaesthete Lorde from composing a song. In a Q&A on Tumblr, the New Zealand singer-songwriter revealed: 'If a song's colours are too oppressive or ugly, sometimes I won't want to work on it. When we first started playing "Tennis Court" [a track on her debut album *Pure Heroine*], we just had that pad playing the chords and it was the worst textured tan colour and it made me feel sick. When we figured out that pre-chorus and started the lyric, the song changed to all these incredible greens.'

❖ ❖ ❖

'Marlo Venus was a beautiful lass / she had the world in the palm of her hand / but she lost both her arms in a wrestling match / to win a brown-eyed handsome man.'
'Brown Eyed Handsome Man', Chuck Berry, 1956

Who was the brown eyed handsome man honoured by Chuck Berry? If you fast-forward to the last verse, it becomes clear that the hero who won the game with a 'high fly hit into the stand' is Jackie Robinson, the first African-American to become a Major League baseball star. A baseball aficionado, and fervent supporter of the St Louis Cardinals, Berry was 22 when Robinson hit a home run, triple, double and single for the Brooklyn Dodgers against his home-town team in August 1948.

More broadly, 'brown eyed' is Berry's code for brown skinned, and his lyric is an oblique allusion to the effect that black American men had on some white women. The first verse, however, is said to have been inspired by Berry's observing a Los Angeles cop trying to arrest a Hispanic man for loitering but being talked out of it by an irate woman: 'The judge's wife called up the district attorney / She said "Free that brown-eyed man" / If you want your job, you better free that brown eyed man.'

Though not one of Berry's biggest hits – it reached #5 on the Billboard rhythm and blues chart in 1956 – the song has been covered by the likes of Buddy Holly, Paul McCartney, Robert Cray, Waylon Jennings and, in one of the most joyous collaborations in the history of rock and roll, by the 'Million Dollar Quartet' – Elvis Presley, Jerry Lee Lewis, Carl Perkins and Johnny Cash – at Sun Studios in Memphis on 4 December 1956. Their exuberant rendition is surpassed only by Nina Simone, whose spirited cover, recorded in 1967, is both a great pop performance and an easy-on-the-ear plea for racial equality.

'We might have a few odd limbs around somewhere. But not enough to make any more paint. We sold our last complete mummy some years ago for, I think, £3. Perhaps we shouldn't have. We certainly can't get any more.'
Geoffrey Roberson-Park, managing director of London colour makers
C. Roberson, 1964

Made by mixing the desiccated flesh of Egyptian mummies with white pitch and myrrh, 'Mummy brown' was one of the most ghoulish pigments ever to disgrace an artist's palette. The resulting pigment ranged in hue from greeny-brown raw umber to ruddy burnt umber.

Some painters – such as Eugène Delacroix, who probably used it in one of his most famous works, *Liberty Leading the People* (1830) – knew exactly how the pigment was made. Others assumed the name merely described a specific shade of brown. The Pre-Raphaelite artist Edward Burne-Jones was one such innocent. One Sunday, as his widow Georgina recalled 'We were sitting together after lunch … the men talking about different colours that they used, when Mr Tadema [the artist Lawrence

Martin Drolling's *Interior of a Kitchen*, a painting with royal heritage.

Alma-Tadema] startled us by saying he had lately been invited to see a mummy in his colour man's workshop before it was ground down into paint. Edward scouted [scornfully rejected] the idea … but when assured it was actually compounded of real mummy, he left us at once, hastened to the studio, and returning with the only tube he had, insisted on our giving it a decent burial there and then.'

In medieval Europe, mummies were imported from Egypt in the mistaken belief that they could cure the sick. Abd al-Latif al-Baghdadi, a twelfth-century physician, insisted that the substance 'found in the hollows of corpses in Egypt' was so similar to bitumen (or, as we know it, asphalt), it could be used in its stead as a medicine. By the sixteenth and seventeenth centuries, 'mumia' – which could be applied to the injured body part or swallowed in a drink – was being dispensed by apothecaries in Europe to cure everything from epilepsy to fractured limbs.

The trade in 'mumia' became so lucrative that, visiting Alexandria in 1564, the French physician Guy de la Fontaine discovered that merchants were taking fresh corpses – often of criminals or slaves – treating them with bitumen and leaving them in the sun to blacken, so they would look ancient. Describing a visit to Memphis, Egypt, in 1586, London merchant John Sanderson noted: 'We walked upon the bodies of all sorts and sizes ... I broke all the parts of the bodies to see how the flesh was turned to drugge and brought home divers heads, hands, armes and feet, for a shew [show].' The medicinal use of mummies waned in the eighteenth century, when the public finally realised that 'mumia' cured nothing.

Artists persevered with such pigments as Mummy brown, Caput Mortuum (literally 'dead head' in medieval Latin) and Egyptian brown, even though these colours tended to fade quickly. Art conservator and historian Sally Woodcock suggests, in her article 'Body Colour: The Misuse of Mummy', that 'It is possible that it continued to be popular because it was easy to handle, gave an initially pleasing effect and artists were unable to resist the allure of painting with a material of such antiquity.' It was rumoured that Martin Drolling's *Interior of a Kitchen* (1815), which hangs in the Louvre, made use of a pigment produced from the remains of French kings disinterred from the abbey of Saint-Denis. Pre-Raphaelite artists such as Tadema, who often painted Egyptian scenes, may well, as Woodcock says, have used mummies to paint mummies. By the end of the nineteenth century the pigment was falling out of favour, and artists were instead using synthetic equivalents. Even so, in 1904 colour makers, Robersons advertised in the *Daily Mail*: 'We require mummy for making colour. Surely a 2,000-year-old mummy of an Egyptian monarch can be used for adorning a noble fresco without giving offence to the ghost of the departed gentleman or his descendants.'

THE BLACK STUFF

Is black really monochromatic? Hokusai, the great Japanese draughtsman, printer and painter, eloquently made the point that black is not a simple thing. 'There is a black which is old and a black which is fresh. Lustrous black and matt black, black in sunlight and black in shadow,' he noted. 'For the old black one must use an admixture of blue, for the matt black an admixture of white, for the lustrous black gum must be added. Black in sunlight must have grey reflections.'

Deep black comes in many varieties, such as pitch (a petroleum-derived resin), jet (a gemstone created from wood under extreme pressure), black spinel (another gemstone), peach stone black (carbonised peach stones), lampblack (soot), vine black (charred vine leaves and twigs), charcoal (from wood burned with minimal oxygen), bone black (animal bones) and obsidian (volcanic glass). None of these, however, are as thoroughly black

Redemption of Vanity – a work created by MIT artist in residence Diemut Strebe in which a carbon-nanotube-coated diamond disappears in a black void.

as the coating made from carbon nanotubes, invented at the Massachusetts Institute of Technology in 2019. Scientists claim this ultimate black absorbs 99.995 per cent of incoming light, making it even darker than the Vantablack developed by British company Surrey NanoSystems in 2016 (see *Pink*). The MIT team, led by Brian Wardle, Professor of Aeronautics and Astronautics, created the coating while experimenting with ways to grow carbon nanotubes on aluminium to improve its conductivity. The tiny, forest-like clusters of carbon nanotubes trap almost all incoming light, so the material appears as a black void.

Black holes – places in space where gravity is so strong that nothing, not even light, can escape – are invisible even to astrophysicists, who deduce their location by observing their impact on neighbouring objects. There are three kinds of black hole: small primordial ones, thought to have been formed soon after the Big Bang; larger stellar ones, created when the centre of a massive star collapses in on itself; and supermassive ones, which are believed to have formed at the same time as their surrounding galaxy. Sagittarius A*, a supermassive black hole at the centre of the Milky Way, has a mass equivalent to four million suns.

Which direction is black? That sounds like a strange question, but the Chinese, Navajo and Turkish would agree that it was north. Many civilisations have assigned colours to the cardinal directions. To the Maya, north was white and black was west. The Turkish name for the Black Sea – *Karadeniz* – might signify its location to the north of Turkey, but it's also possible that the name comes from darkness of the water, caused by a high level of iron sulphide. We know that, in around 500 BCE, the Achaemenids who ruled Persia described the sea as *axsaina* (Persian for 'dark'), a term which was mistranslated by Greek settlers as *axeinos* ('inhospitable'), which then became *euxinos* ('hospitable').

Philip Melanchthon (1497–1560), theologian of the Reformation, friend and colleague of Martin Luther, was aptly named: 'Melanchthon' is a classicised version of *Schwarzterdt*, meaning 'black earth'. He nursed a visceral hatred of bright colours, fulminating about fashions that clothed 'men like peacocks'. True Protestants, he proclaimed, never wore anything flamboyant. Black signified humility, contrition and awareness of sin, whereas bright colours reeked of Catholic corruption. This insistence on the moral virtue of a monochromatic colour scheme had disastrous consequences for many church buildings in Switzerland, the Netherlands, England and southern Germany, where Protestant zealots destroyed many of the stained-glass windows that were one of the glories of the Middle Ages.

The black robes of Protestantism were often fabricated using logwood, a natural dye recently discovered by Spanish colonialists in Central America. When Hernán Cortés came to Mexico in 1519, he was struck by the vivid black and violet dyes that the Aztecs used. The critical ingredient in these dyes, Cortés discovered, was the dark red heartwood of logwood trees, a little of which would, when chipped, soaked, boiled, drained and fixed with a mordant, produce copious amounts of dye. Using iron as the mordant produced an inexpensive black, a significant advance given that, hitherto, the only means of creating a durable black dye involved triple-dipping cloth in vats of blue, red and yellow, a difficult and expensive process.

The English Parliament banned the use of logwood in 1581, claiming that it produced colours of a 'fugacious character' which faded quickly. There was also an economic consideration, as Spain was the principal beneficiary of the logwood trade – this was only seven years before Philip II dispatched his ill-fated armada. The ban was repealed in 1673, by which time Britain had its own supply. Extracting the heartwood from mosquito-infested mangrove swamps in what is now Belize was arduous and hazardous, but by the mid-1600s the 'bay men', as they were known, were rich enough to establish logging camps and hire slaves to take on some of the hard labour. The 'bay men' worked hard and played hard. The irony, as Victoria Finlay points out, was that the dye by means of which the Puritans advertised their piety was 'was bought from men who spent all their profits on rum, women and partying'.

The goddess Kali – her name, in Sanskrit, means 'she who is black' or 'she who is death' – emerged in South Asia in the sixth century CE, and is the most terrifying deity in the Hindu pantheon, as this description by the American scholar Wendy Doniger makes clear: 'Kali is most often characterised as black or blue, partially or completely naked, with a long lolling tongue, multiple arms, a skirt or girdle of human arms, a necklace of decapitated heads and a decapitated head in one of her hands.' With her many arms, and her penchant for dancing or standing on her prostrate husband Shiva, the black goddess was briefly reinvented as a feminist icon, appearing on the front cover of the first issue of the American magazine *Ms.* in 1972.

On an episode of his weekly chat show in 1971, shortly after recording his own protest song 'The Man in Black', Johnny Cash

addressed the issue of his monochrome wardrobe. 'I've worn black basically since I was in the music business,' he told interviewer Mike Douglas. According to the lyrics of his song, he wore black to highlight all kinds

of human suffering, concluding, 'I'd love to wear a rainbow every day / And tell the world that everything's okay', but 'until things are brighter I'm the Man in Black.' *Rolling Stone* magazine has a more prosaic explanation for Cash's sartorial preferences: 'He took to black because it was easier to keep clean on long tours. Early in his career, fellow acts teased him about it, calling him "The Undertaker".'

❖ ❖ ❖

It's strange that nobody ever applied Cash's nickname – or anything like it – to Roy Orbison, whose black clothes, black-dyed hair and dark glasses became his trademark. 'I wasn't trying to be weird, I didn't have a manager who told me how to dress or present myself, but the image developed of a man of mystery, a quiet man in black,' he explained. That image suited such heartbreaking ballads as 'It's Over', 'In Dreams' and 'Crying', which he sang like a man in the depths of mourning. For much of the late 1960s and early 1970s, that is exactly what Orbison was. His first wife, Claudette, whom he had divorced because of her infidelities but then remarried, was killed in a motorcycle accident in 1964. In September 1968, his home in Hendersonville, Tennessee, burned down, killing his two eldest sons. Orbison's friend and neighbour Johnny Cash bought the house, demolished it and turned the land into an orchard.

The black aces of spades and clubs were implicated in the killing of 'Wild Bill' Hickok on 2 August 1876. Hickok was playing five-card stud poker in a Deadwood saloon when a drunken gambler called Jack 'Broken Nose' McCall shot him, point blank, in the back of the head. Hickok was holding two black aces and two black eights, plus another unknown card, which was spattered with his blood. Even though the expression 'dead man's hand' was already in use for entirely different combinations of cards, a 1926 biography of 'Wild Bill' incorporated the two aces and two eights into his legend and made them definitive.

In old-style Hollywood swashbucklers, pirates always sail under the Jolly Roger, a black flag with white skull and crossbones in the centre. For once, the dream factory got it right: the French pirate Emanuel Wynn is often said to have been the first to fly the Jolly Roger, roughly around 1700. Contemporaneous brigands such as Blackbeard and Charles Vane used the same device. Although other pirates sailed under different colours, the black flag remained the most popular – it was a famously fearsome sight that acted as a coded message to intended victims, giving them a chance to surrender or abandon ship. If they spurned

The earliest known Jolly Roger: a 1725 woodcut of Stede Bonnet in Charles Johnson's *A General History of the Pyrates.*

that chance, the black flag was lowered and a red flag hoisted, proclaiming their intention to spill blood.

In Robert Louis Stevenson's novel *Treasure Island* (1882), the black spot is a card that is usually presented to a pirate who is about to be killed, overthrown or both. Stevenson describes it in some detail: 'It was around about the size of a crown piece. One side was blank, for it had been the last leaf; the other contained a verse or two of Revelation – these words among the rest which struck sharply home in my mind: "Without are dogs and murderers." The printed side had been blackened with wood ash, which already began to come off and soil my fingers; on the blank side had been written with the same material the one word "depposed". The grisly device may have been inspired by tales of Caribbean pirates showing the ace of spades to traitors and informers before executing them.

'When the little black dress is right, there is nothing else to wear in its place.'
Wallis Simpson, Duchess of Windsor

On the evening of 29 June 1994, before she set off for *Vanity Fair*'s annual fundraiser for London's Serpentine Gallery, Princess Diana changed her mind about the dress she would wear. Annoyed that her original choice had been leaked to the media, she slipped on a figure-hugging black cocktail dress created for her by Greek designer Christina Stambolian. Diana had previously considered the dress 'too daring'. Yet on that particular night she knew that Prince Charles, her estranged

The 'revenge dress': Princess Diana steps out in Stambolian.

husband, was about to admit publicly his adultery with Camilla Parker Bowles, so she opted for the protocol-busting, off-the-shoulder, black outfit. Now known as the 'revenge dress', it became instantly iconic.

The story of the 'revenge dress' begins in south-western France in 1895, when the mother of three young sisters – Julie, Gabrielle and Antoinette Chanel – died. Their nomadic ne'er-do-well father sent them to an orphanage attached to the Cistercian abbey of Aubazine. Forced at the age of twelve to adapt to the austere routine of the Aubazine nuns, Gabrielle learned to sew in a world of monochrome religious habits and school uniforms. This training would, legend has it, inspire Coco Chanel – as Gabrielle became known – to design her little black dress (aka LBD), which, when it appeared on the front cover of *Vogue* in 1926, was hailed as the 'Model T Ford of fashion'.

The LBD revolutionised women's fashion. As Karen van Godtsenhoven, associate curator of the Costume Institute at the Metropolitan Museum of Art, New York, observed: 'Black has an intellectual and cerebral value, and a contradictory ambiguity – it's the colour of power, but also the colour of rebellion. In the 1950s, black was worn by the Beat poets and by Juliette Gréco, the muse of existentialism.' The

Left Bank's most seductive chanteuse, lover of Miles Davis and role model for millions of French women, Gréco looked cooler in black than anyone else, cooler even than Audrey Hepburn in *Breakfast at Tiffany's*.

In the mid-1940s, such film noir heroines as Rita Hayworth (*Gilda*), Lauren Bacall (*The Big Sleep*) and Ava Gardner (*The Killers*) all wore black dresses and yet, despite the Duchess of Windsor's ringing endorsement, the look has never been quite as popular in America as in Europe. 'American fashion has a cultural preference for colours which evoke freshness, vivacity, prettiness,' said van Godtsenhoven. 'Black is a European, French colour, reserved for seductresses who are chic, smouldering and cerebral.'

'Black is modest and arrogant at the same time. Black is lazy and mysterious. Above all black says this: I don't bother you – don't bother me.'
Yohji Yamamoto

Described as 'fashion's poet of black' by the *New York Times*, Yohji Yamamoto is revered by fashionistas for his dedication to creating perfectly cut clothes of a single colour. His love of black is both cultural – the advent of punk in the late 1970s was an inspiration – and personal. His father, an unwilling conscript in the Japanese army, was killed in the Second World War, when Yohji was an infant. From then on, his widowed mother wore 'nothing but mourning clothes'. His mother was a dressmaker, who made floral outfits influenced by European designers, but Yohji became intrigued by black, sensing its potential to empower women. As he said once: 'I make clothing like armour. My clothing protects you from unwelcome eyes.' If that is your

Yohji Yamamoto, armed in black.

goal, black is the logical colour to employ – especially if, by doing so, you are also honouring your lost father.

In his book *The Non-Objective World*, Kazimir Malevich wrote: 'In the year 1913, trying desperately to free art from the dead weight of the real world, I took refuge in the form of the square.' Two years later, he created *Black Square*, the defining work of the movement known as Suprematism, and arguably the starting point for abstract art. His aim, he said, was to 'kill the art of the picturesque and put it in a coffin'. His first black square, however, was designed for the stage curtain of the experimental opera *Victory Over the Sun*, in which the characters seek to abolish reason by capturing the sun and destroying time. He and his collaborators – the poets Aleksei Kruchenykh and Velimir

Malevich's *Black Square* exhibited at The Last Futurist Exhibition of Paintings *0,10*, which introduced Suprematism, Petrograd, 1915–16.

Khlebnikov and musician Mikhail Matyushin – regarded the rejection of rational thought as the prerequisite for revolutionary change. An admirer of Russian mystic P.D. Ouspensky, Malevich claimed his art was divinely inspired, and that he received guidance from cosmic voices.

The *Black Square* was created amid the intellectual, social and political ferment of pre-revolutionary Russia. Lenin was never fond of Suprematism – it was too remote from reality for his tastes – but the Trotskyists embraced it and it briefly became the Bolsheviks' preferred visual idiom. As Trotsky's power waned, so did Malevich's reputation. (He smuggled some of his most avant-garde work out of the Soviet Union in the 1920s.) Stalin's diktat that Social Realism was the artistic genre most likely to 'safeguard the interests of the working class' effectively forced Malevich to change his style, but he quietly protested against the new orthodoxy by signing many of his more conventional landscapes with a small black square. When he died in 1935, at the age of 56, a black square was painted on his coffin.

In 2015, in preparation for the centenary of Malevich's *Black Square*, an X-ray of the earliest version of it (he painted four) revealed a piece of text underneath the paint: 'Negroes battling in a cave.' This is almost certainly an allusion to a satirical print by Alphonse Allais of a completely black canvas, subtitled *Negroes Fighting in a Cellar at Night*, which in turn was a reference to an entirely black painting with a very similar title by his friend Paul Bilhaud. More than a century later, such jokes – like Malevich's text – are likely to strike us as racist rather than amusing, but Allais's print was one of a series of monochrome visual puns, including the entirely white *First Communion of Anemic Young Girls in the Snow* and the all-red *Apoplectic Cardinals Harvesting Tomatoes on the Shore of the Red Sea*.

French artist Pierre Soulages (1919–) rarely uses any colour other than black, a hue that for him signifies the beginning of art, in the darkness of ancient caves. His obsession with blackness began at an early age – the six-year-old Soulages was once found by his older sister drawing thick black lines in ink with a brush, and when she asked what he was sketching, he replied: 'Snow'. He calls his abstract style *outrénoir* – 'beyond black' – to indicate that he doesn't work with black but with the light that black reflects.

Why do we get ill? Thousands of years ago, the answer to that question was simple, obvious and consistent: no matter what the symptoms were, if we were sick, we were possessed by bad spirits. Medical diagnosis became a bit more nuanced in ancient Greece when Hippocrates (460– 377 BCE) and his fellow physicians suggested that many diseases had natural causes, not supernatural ones.

They developed the idea that our bodies had four humours (or fluid substances) which matched the four elements. Blood was hot and wet like the air, yellow bile was hot and dry like fire, phlegm was cold and wet like water and black bile was cold and dry like the soil. Getting sick was a sign that our bodies generated too much or too little of one or more of these humours. It was the doctor's job to find out which humour was out of balance and treat it – which, in ancient Greece, often meant extracting it (by letting blood) or changing the patient's diet.

Dry skin, vomiting, depression were regarded as signs that the patient was suffering from too much black bile. The word 'melancholy' comes from the Greek *melas* or *melan* ('black') combined with *khole* ('bile').

'Your trouble – I mean the Black Dog business – you got from your forebears. You have fought against it all your life'.
Lord Moran, Winston Churchill's physician, addressing his patient.

Winston Churchill's beloved nanny, Elizabeth Ann Everest, introduced him to the image of a 'black dog' when describing his dark moods as a child. Everest understood Winston's

From Kaye Blegvad's graphic novel about depression, *Dog Years*.

abject loneliness as a boy in a way that his mother and father Randolph and Jennie, distant and neglectful even by the dismal parental standards of the Victorian aristocracy, never did. (In fairness, Randolph Churchill probably suffered from bipolar depression, too.) English psychiatrist Anthony Storr argued that, given the severity of his depression, Churchill's defiance of the 'black dog' was heroic. Only after resigning from office in 1955 did Churchill finally succumb: he would sit for hours, without speaking or reading, as, in his daughter Mary's words, 'a cloud of black despair' descended.

Where does the black dog metaphor come from? An essay by Paul Foley concludes that black dogs did not, as is often suggested, become synonymous with depression in Roman times. Horace's oft-quoted phrase 'the black dog follows you' could, he argues, be a mistranslation of 'that dark companion dogs your flight'. Foley suggests that the image arose in the eighteenth century.

The notoriously depressive Samuel Johnson (1709–84) once complained that he shared every solitary breakfast with a black dog. The fact that he also wrote to a friend, 'What will you do to keep away the black dog that worries you at home?' suggests this might already have been a popular idiom.

At the height of the Industrial Revolution, the region of England's West Midlands known as the Black Country – which is generally accepted to comprise most of Dudley, Sandwell, Walsall and Wolverhampton – was very, very black indeed. 'The country is very desolate everywhere. There are coals about and the grass is quite blasted and black,' the 13-year-old Princess Victoria noted in her journal, when she visited the Black Country in 1832. 'The men, women, children, country and houses are all black.'

To understand young Victoria's horror, you only have to gaze at *Black Country, Night, with Foundry*, a painting by Edwin Butler-Bayliss which brilliantly evokes William Blake's 'dark, satanic mills'. An ironmonger's son, who grew up in Wolverhampton, Butler-Bayliss was painting what he saw. In the 1860s, Elihu Burritt, the American consul to Birmingham, described the region as 'black by day and red by night [and] cannot be matched for vast and varied production by any other space of equal radius on the surface of the globe'.

The Black Country was black even before it was industrialised – the name is linked to the South Staffordshire coal seam, also known as the '30ft seam', said to be the thickest layer of coal in Britain, which darkens the soil – but the soot and smoke from the thousands of factories made it blacker still, and shortened people's lives. In 1841, the parish of Dudley had the worst mortality rate in Britain.

Black Country, Night, with Foundry by Edwin Butler-Bayliss. Its subject, Dudley, may have been the inspiration for Tolkien's Mordor.

You can understand why some scholars argue that the Black Country was the model for Mordor, the desolate land invented by J.R.R. Tolkien (who grew up in Hall Green, a suburb of Birmingham) for *Lord of the Rings*. In Tolkien's elvish Sindarin language, 'Mordor' means 'Black (or Dark) Land'.

The Black Country is greener than it used to be. The foundries and factories that once blackened the sky are now gone, and the last local coal mine, Baggeridge Colliery near Sedgeley, which closed in 1968, is now a country park. While welcoming a healthier environment, some locals worry that the Black Country is losing its identity. Exploring Mordor's origins for the *Guardian* in 2014, author Stuart Jeffries, who was raised in Dudley, recalled the 'terrible, intoxicating beauty of the Black Country ablaze at night, the fumes that caught in the back of your throat and made you know you were home'. That sense of loss has been sharpened

by the destruction of the area's high streets, but the regional identity lives on in the local accent, dialect and vocabulary.

That 'fumes in the back of your throat' feeling is still available
elsewhere in the world, especially in Kuzbass, the part of Siberia
where more than half of Russia's coal is mined. In the winter of
2018, the air in some towns was so thick with soot and ash that
toxic black snow fell from the sky. Local environmentalist Vladimir
Slivyak complained: 'It's harder to find white snow than black
snow.' The authorities blamed a faulty shield that had failed to
trap the dust from a massive coal plant. In Myski, home to around
45,000 residents, the authorities had the black snow painted white.

When the very first rubber car tyres were invented in 1895, they
were white. Soot was soon added because it was thought to make
them last longer, but, around 1910, companies began using the
chemical compound carbon black as a tyre filler. Carbon black
strengthens tyres, protects them from UV light and ozone and
conducts heat away from the tread, which gets hot during a long
journey. It is estimated that around 70 per cent of the world's
carbon black is used to make tyres.

Black Friday is a strange way to describe the shopping bonanza
that follows Thanksgiving in America but, according to the
Oxford English Dictionary: 'The likely story is that Black Friday
started out as a joking reference to how bad the traffic would be
on this day.' Early citations in the *OED* suggest that Philadelphia's bus drivers and police officers coined the phrase, as they

braced themselves for congestion and chaos on the roads. Even in an age of online shopping, the term lingers on.

The Ayam Cemani is a chicken breed that's native to Indonesia, and its name, which means 'thoroughly black chicken', reflects the fact that its feathers, meat, bones and internal organs are all black – a condition caused by a surfeit of the gene EDN3, which causes hyperpigmentation. The Ayam Cemani is rare and revered – it is reputed to have magical powers, and is believed to have been used in religious rituals since the twelfth century.

In Italy, the Black Hand, aka Mano Nera, was the name of a method of extortion that seems to have arisen in the Kingdom of Naples in the middle of the eighteenth century. Threats were delivered in person, by leaving a coal-blackened handprint on the victim's door, or by post – with a letter signed with the image of a black warning hand. By the turn of the twentieth century, criminal gangs known as the Black Hand were using the same techniques to terrify people in various US cities, from San Francisco to Chicago and New York.

The Black Hand is also the name of secret Serbian military society (Crna Ruka) that was formed in 1901 with the aim of uniting all the Slav-majority territories of the Balkans. Historians still argue about the extent to which the Black Hand was involved in the assassination of Archduke Franz Ferdinand in Sarajevo in

Chicago Black Hand note, 1907.

1914, a murder committed by a Bosnian Serb named Gavrilo Princip, which precipitated the outbreak of the First World War.

In Italy, the Blackshirts were the shock troops for Mussolini's brutal dictatorship, and black became the brand of his regime. The choice of colour is often traced back to the Arditi ('Daring Ones'), the Italian special force, resembling the British SAS and the US Navy SEALs, which fought in the Alps during the First World War. The Arditi did wear black, but not exclusively. Mussolini might also have been inspired by the poet, politician, pilot and philanderer Gabriele D'Annunzio, who in September 1919, aided by 2,600 nationalists, seized the Adriatic port of Fiume (now Rijeka in Croatia) and reclaimed it for Italy. He ruled for only fifteen months but it was long enough for Mussolini to study his playbook – the *braggadocio*, the Roman salute, the black-shirted followers – and apply similar techniques to seize power in March 1922.

Appalled by the slovenly appearance of some of Mussolini's supporters during their march on Rome in that year, Paolo Garretto (1903–89), a teenage Fascist, caricaturist and graphic designer, smartened up the uniform. As he recalled: 'I did not like the way they were all dressed up. They had only one common garment – the black shirt. As for the rest of their uniform, they wore anything they liked, such as long trousers of any colour. So, I designed for myself a uniform that was all black – shirt, cavalry trousers and boots.' Three friends liked his uniform so much they copied it and called themselves 'the Musketeers'. Spotting them on parade, Mussolini ordered them to join his honour guard.

'The black shirt is not the everyday shirt, and it is not a uniform either. It is a combat outfit and can only be worn by those who harbour a pure soul in their heart,' Mussolini bombastically declared. As Simonetta Falasca-Zamponi noted, in her book *Fascist Spectacle: The Aesthetics of Power in Mussolini's Italy*: 'The uniform, along with the other daily rituals the regime insisted upon, was intended to wipe out bourgeois mentality and habits.' Black shirts were an integral part of Mussolini's campaign to unite Italians around a mystical belief in their national destiny.

Mussolini's sartorial innovation inspired many foreign Fascist movements. Oswald Mosley's British Union of Fascists, the Dutch Black Front and the Nazi SS likewise chose black. In the States, the fascistic Proud Boys' penchant for black Fred Ferry polo shirts prompted the company to halt exports of that particular design to North America. But blue was the colour for Fascists in Canada and Ireland, and the Spanish Falange, whereas the Iron Guard in Romania wore green and the South African Gentile National Socialist Movement grey. Fascist shirts were blingier in Mexico, where members of the Revolutionary Mexicanist Action party, launched by Nicolás Rodríguez Carrasco, were resplendent in eye-catching gold shirts.

In Muslim tradition, a black flag is one of the standards flown by Muhammad. In 747–750, the Abbasids, who claimed descent from the prophet's uncle, al-Abbas, overthrew the Umayyad caliphate in a revolution also known as the Movement of the Men of the Black Raiment. To legitimise their rule, the Abbasids carried black banners which, in Muslim teachings, are flown by the army with which the Mahdi ('rightly guided one' in Arabic) will establish just rule and cleanse the world of evil.

The association with the Mahdi – and with traditions of a battle that prefigures the end of times at the Syrian town of Dabiq – has prompted various modern jihadist movements, including al-Qaeda, al-Shabaab and ISIS, to adopt black flags.

'The black flag is the negation of all flags. It is a negation of nationhood ... Black is a mood of anger and outrage at the hideous crimes against humanity perpetrated in the name of one state or another ... But black is also beautiful. It is a colour of determination, of resolve, of strength, a colour by which all others are clarified and defined.'
Howard J. Erlich, Reinventing Anarchy, Again (1996)

Black, as a kind of no-colour, seems a logical hue for anarchists to adopt – Peter Kropotkin, one of the leading proponents of

The arrest of the anarchist Communard Louise Michel, May 1871, by Jules Girardet.

anarchy, described it as the 'no-government form of Socialism'. Yet the origins of the black flag of anarchism are far from clear. Black flags were flown by striking silk workers in Lyon in 1831 as they chanted 'Bread or lead' to signify their willingness to die for their cause. The strikers, who took over the city for a time, were not anarchists as such, but their actions would have been known to the philosopher Pierre-Joseph Proudhon, who stayed in Lyon in 1843–44, and is the first person known to declare himself an anarchist. There's a clearer connection with the 20,000 Communards who were massacred by the French Army in 1871. One of the surviving Communards, Louise Michel, flourished a black flag at a demonstration in Paris in 1883, calling it 'the flag of strikes and of those who are hungry'.

Niger, the Latin word for 'black', gave rise to the French *noir*, was used to name countries (Nigeria, Niger) and rivers (Niger), metamorphosed into *negro* ('black' in Spanish), and mutated, in the US and UK, to the racial insult 'nigger'. In America, the N-word has also served as an all-purpose term for any foreigners who didn't have white skin. At the turn of the twentieth century, as the American military suppressed the Philippines independence movement, one soldier wrote home: 'Our fighting blood was up and we all wanted to kill niggers.'

The changing language of ethnic blackness is reflected in the racial categories used in the American census. In 1790, people were asked to identify themselves as a free white male, free white female, other free person or slave. Thirty years later, the category 'colored' was added. In 1890, black Americans were asked to identify themselves as black, mulatto (if they were half-black), quadroon (one quarter black) and octoroon (one eighth). We don't know how many citizens belonged to each category, because most of the census records were destroyed by fire.

In 1906, when Booker T. Washington, the black author, academic and presidential adviser, was asked by *Harper's Weekly* to suggest the right term for 'coloured' people, he replied that 'it had long been his own practice to write and speak of members of his race as negroes and, when using the term as a race designation, to employ the capital N'.

Three years later, the organisation we now call the National Association for the Advancement of Colored People was founded in New York. It was initially known as the National Negro Committee, but 'Colored' was subsequently preferred because, at the time, it was deemed the polite way to refer to black people. In 1930, 'Negro', with a capital 'N', made it into the *New York Times'* style guide. In that year's census, anybody with any black ancestry was requested to identify themselves as Negro. Yet by the 1960s, many felt that the term implied subservience to white people. As Ossie Davis, the actor and civil rights activist, told *Ebony* magazine: 'Malcolm X used to be a negro but he stopped. He no longer depended on white folks to supply his needs – psychologically or sociologically – to give him money or lead his fight for freedom or protect him from his enemies, or to tell him what to do.'

Black Panther poster, 1969.

On 16 June 1966, the activist Stokely Carmichael declared in a speech in Greenwood, Mississippi: 'The only way we gonna stop them white men from whuppin' us is to take over. What we're gonna start sayin' now is Black Power.'

George Floyd Memorial in Portland, Oregon, June 2020.

Three months later, the Black Panther Party was founded in Oakland, California.

Carmichael's stance was challenged in December 1988 by the Reverend Jesse Jackson who declared, 'To be black is baseless. Black does not describe our situation. We are of African-American heritage.' The term 'African-American', he argued, had 'cultural integrity' and put black Americans in their 'proper historical context'. By 2003, a poll showed that the hyphenated term was preferred by nearly half of 'black' Americans.

One of those who didn't prefer it was singer-songwriter Smokey Robinson, who observed: 'If you go to Africa in search of your race, you'll find out quick you're not an African-American, you're just a black American taking up space.' In his view, the hyphenated term should only apply to people who have come to America from Africa in the past twenty or thirty years. Some recent immigrants call themselves Continental Africans on the grounds that they were born and bred in Africa and hope one day to return there.

The two terms have been used almost interchangeably for decades, but the term 'Black' with a capital 'B' found increasing favour after the killing of George Floyd in police custody in Minneapolis on 25 May 2020. Several news publishers – notably Associated Press, the *New York Times* and *USA Today* – changed their style guides to give 'Black' an upper-case 'B'.

On 2 June 2020, more than 28 million people posted a plain black square with the hashtag #blackouttuesday on social media platforms (mainly Instagram) and closed their accounts for the rest of the day. The global initiative was in protest against George Floyd's death in police custody and in support of the Black Lives Matter movement, which had been founded seven years before.

BLM was launched by three black activists – Alicia Garza, Patrisse Cullors and Opal Tometi – in response to the acquittal in 2013 of George Zimmerman, who claimed he had acted in self-defence when he shot and killed Trayvon Martin, a seventeen-year-old African-American, in Sanford, Florida. Between 2015 and May 2020, according to a study for the *Journal of Epidemiology & Community Health*, 27 per cent of the 4,653 deaths Americans shot and killed by police were black – a shocking statistic given that black Americans account for 13 per cent of the population.

THE GREY AREA

G rey is associated with many of our least favourite things: miserable weather, growing old, fog, corporate conformity and monochrome urban environments rendered dismal by concrete and cement. Arriving from New Zealand in 1946, at the age of fourteen, Franklin Birkinshaw (better known as novelist Fay Weldon), wrote: 'Was this my mother's promised land? Where were the green fields, rippling brooks and church towers? Could this be the land of Strawberry Fair and sweet nightingales? Here was a grey harbour and a grey hillside, shrouded in a kind of murky, badly woven cloth, which as the day grew lighter proved to be a mass of tiny dirty houses pressed up against one another, with holes gaping where bombs had fallen, as ragged as holes in the heels of lisle stockings. I could not believe that people actually chose to live like this.' When her mother told her that this was just how the Essex port of Tilbury looked, she wondered: 'Just Tilbury? The greyness was vast, as far as the eye could reach.'

In *The Luminous and the Grey*, David Batchelor lists 48 synonyms for the colour from the *Collins English Thesaurus*, and the most positive – in a selection dominated by variants of 'depressing', 'gloomy', 'dismal' and 'nondescript' – are 'mature' and 'neutral'. With all this going on, E.L. James was brave to call her erotic bestseller *Fifty Shades of Grey* – even if the Grey in the title is the name of one of the book's protagonists.

Grey's dour, dismal image harks back to at least the early years of the thirteenth century, when Icelandic scholar Snorri Sturluson is said to have written and compiled the *Prose Edda*, one of the key texts of Old Norse literature. In one famous tale, a gigantic serpent is described as *grár*, an Icelandic word which, according to etymologist Anatoly Liberman, describes the colour grey but also means 'spiteful', 'evil' and 'hostile'. 'We don't know whether Sturluson relished the pun or even meant the adjective to be understood in both senses,' Liberman noted, pointing out that wolves were also called *grár* because they were grey and feared.

The same pun can be found in German: *grau* is 'grey', but add the suffix *–sam* and one gets *grausam*, which means 'cruel' or 'terrible'. *Grausam* is, Liberman observed, pretty close to 'greysome', an adjective meaning 'grisly' or 'horrific', which in its modern spelling is 'gruesome', a word that entered common parlance through the novels of Walter Scott.

The rarity of grey eyes – around one per cent of the world's population have them – has prompted various superstitions and beliefs. They have been taken to signify greed (according to an

Megan Fox – a grey-eyed Teenage Ninja Turtle.

old New England nursery rhyme), wisdom (Athena, goddess of war, peace and wisdom, had grey eyes) and good marksmanship (Ambrose Bierce's short story 'An Occurrence at Owl Creek Bridge' includes the line 'He observed that it was a grey eye and remembered having read that grey eyes were keenest, and that all famous marksmen had them'.

In the Arctic city of Murmansk, in December and January, a grey Saturday afternoon is something to look forward to. From 2 December until 11 January every year, the sun does not rise above the horizon and for most of this period the closest the city comes to daylight is an hour or two of grey twilight around noon. Most locals become accustomed to this – although some struggle with depression, fatigue and insomnia every winter – but visitors can find this greyness more unsettling than the dark. It feels as if you have wandered into a post-apocalyptic movie. On the city's outskirts, this greyness is amplified by the concrete blocks of Soviet-style apartments but, in the centre, the gloom is interrupted by buildings painted in pastel pink, green and orange. Nature does its best to help, disrupting the long dark polar night with the aurora borealis.

'For many people, grey is intrinsically unappealing, uninteresting and unexciting – a sort of Cinderella of the artist's palette which never gets to the Technicolor ball … There is no grey in the rainbow, Mother Nature's most dazzling and bravura display of light and warmth, and thus of joy and hope.'
David Cannadine, president of the Royal Society

Cannadine has a point, but grey has been deployed to powerful effect by several great artists, such as Caspar David Friedrich and, most famously, James McNeill Whistler, whose best-known painting is his *Arrangement in Grey and Black No. 1* (1871), more widely known as *Whistler's Mother*. Claude Debussy, an admirer of Whistler, described his *Nocturnes* (1899) as 'an experiment in the different combinations that can be obtained from one colour – what a study in grey would be in painting'. Claude Monet's studies of the Thames, often undertaken on foggy, dismal days, were thoroughly grey.

Whistler's *Arrangement in Grey and Black No. 1.*

One of the colour's charms is that, as David Batchelor writes, 'It is close to impossible in practice to find a grey that is not inflected by some other colour, although the not-grey of grey often becomes more visible only as two or more greys are placed next to each other.' German artist Gerhard Richter (1932–) makes a similar point: 'When I first painted a number of canvases grey all over ... I did so because I did not know what to paint or what there might be to paint ... As time went on, however, I observed differences of quality among the grey surfaces – and also that these betrayed nothing of the destructive motivation that lay behind them. The pictures began to teach me.'

The most powerful use of grey in modern art is Pablo Picasso's *Guernica*. The artist read about the Luftwaffe's carpet-bombing of the Basque city on 26 April 1937 in a newspaper in Paris, where he was working on a mural for the Spanish government, for that summer's Paris exhibition. Having read the horrific description of the raid, which left 1,654 people dead and 889 wounded, Picasso abandoned the mural and set to work on *Guernica*. The first iterations were in colour, but Picasso's artist friends argued that this diminished its power. It's possible, too, that a series of black and white photographs taken of him at work on the painting may have influenced the decision to limit his palette to the tones of newsprint. The finished work, in the words of critic Herbert Read, is a 'modern Calvary'.

'No light here penetrates through common glass and the effect is magical; the superb rose and lance windows, not dazzling, rather captivating the vision with the hues of the rainbow, being made up, as it seems, with no meaner materials than sapphire, emerald, ruby, topaz, amethyst, all these in richest, unimaginable profusion; other interiors are more magnificent in architectural display, none are lovelier than this.' This passage, from an article published in *Fraser's Magazine* in 1878, brilliantly evokes the mesmerising effect Troyes Cathedral's 180 stained-glass windows have had on millions of visitors since the thirteenth century.

To the Cistercians, an austere order of monks and nuns founded in 1098, such flamboyance was a distraction from the worship of God. Bernard of Clairvaux, the Burgundian abbot and mystic who became the predominant influence in the new order, loathed imagery of all kinds, grandiose architecture, and colour: 'For God's sake, if men are not ashamed of these follies, why at least do they not shrink from the expense?' The interiors of the first Cistercian abbeys were accordingly austere and monochrome. When glass was used, it was white or stained grey, a colour known as grisaille, after *gris*, the French word for 'grey'.

Grisaille has been used as an underpaint layer by many artists, but some have also used it for trompe l'oeil effects. Completed in the early fourteenth century, Giotto's allegorical frescoes of the seven virtues and vices, which adorn the north and south walls of the Scrovegni Chapel in Padua, have deceived many visitors into thinking the images are three-dimensional. Henry Moore described them as 'the finest sculpture I met in Italy'. It was not Giotto's only experiment with deception. As Giorgio Vasari recalls in *Lives of the Artists*: 'One day, Giotto decided to play a

Hope and Inconstancy from Giotto's allegorical frescoes at Padua's Scrovegni Chapel.

trick on the older artist Cimabue, to whom he was apprenticed. So when the latter's back was turned Giotto painted a tiny fly onto the mural which his master was painting. Cimabue went berserk trying to brush away the fly, before he realised it was an illusion.'

❖ ❖ ❖

'Rocks, having colour, mountains (remote), mountains (nearer), clouds, cold.'
Landscape painter John Ivey's recommended uses of Payne's grey

Although Payne's grey is perennially popular with artists and musicians, few people remember William Payne (1760–1830), the English water-colourist who first concocted it. At some point – no one really knows when – he invented a neutral tint of grey to help him capture 'atmospheric perspective' – the way hills and mountains look paler and bluer the further away they are. There was no set formula for the original Payne's grey, because he mixed it as and when required, but blogger Katy Keller describes the basic mix as a 'good deal of Prussian blue, a smidge of yellow ochre, and a dab of crimson lake'. Whereas Payne's own grey was made with three primaries, paints sold under that name today are generally a simple mixture of a black and a blue pigment. Many artists find the modern commercial Payne's grey too dull, and Australian water-colourist and teacher Jane Blundell was so dissatisfied that she invented her own version, using colours such as burnt sienna and ultramarine to create a 'beautiful, harmonious grey' that she has named, inevitably, Jane's grey.

Although many American states have dismantled Confederate statues and removed the rebel flag from public buildings, ordinary citizens can still buy Confederate grey paint from paint makers such as Maison Blanche, Dulux and Benjamin Moore. Indeed, according to the Maison Blanche website: 'Confederate gray is one of our most popular colours.' Some people buy it to paint figures as they recreate battles from the American Civil War. In fact, Confederate jackets were of various colours: cadet grey (a

Confederate grey troops at the Battle of Franklin, 1864.

foggy blue tint of grey), smoky grey or even butternut brown. At the start of the conflict, some Union troops wore grey, too.

We know from eyewitness accounts that pirates used blue-grey to conceal their ships in the Mediterranean in the third century. This may be the first known use of the colour we know as battleship grey. Adopted by Confederate gun runners during the American Civil War – who found that the colour made it easier to hide their craft in coastal fogs – grey became the colour of choice for the French, German, American and British navies between 1890 and 1907. But a large grey vessel on open water still tends to be a conspicuous object, especially when it's belching smoke, and in 1917, with the First World War raging and German U-boats sinking twenty-three British ships in a week, the Royal Navy began to experiment with other colour schemes.

Norman Wilkinson, a painter, keen yachtsman and lieutenant in charge of a minesweeper in the English Channel, had the brilliant, counter-intuitive idea that, instead of camouflaging ships, it made more sense to camouflage their intent. They could be painted 'not for low visibility, but in such a way as to break up her form and thus confuse a submarine officer about the course on which she was heading'. By painting strange shapes and curves on ships in complementary colours – typically black and white, green and mauve, orange and blue – Wilkinson could make them appear smaller, back to front or to be heading in the opposite direction. The technique – soon known as 'dazzle' – worked best when seen from the low vantage point of a U-boat gunner.

In October 1917, a demonstration with a tiny model ship and a periscope convinced King George V, who had served in the Royal Navy, that the vessel was heading 'south by west' when it

Dazzle camouflage applied to bathing costumes, 1917.

was really heading 'east-south-east'. At sea, the results were less spectacular. An Admiralty study found that, in the first quarter of 1918, 72 per cent of dazzle ships attacked were sunk or damaged, compared to 62 per cent of ships painted in traditional colours. In the next three months, the results were reversed: 60 per cent of attacks on dazzle ships led to sinking or damage, compared to 68 per cent on non-dazzled vessels. The Admiralty concluded that dazzle may not have been a great help, but it probably did no harm either, especially because sailors and insurers believed in its effectiveness, which meant higher morale on – and lower premiums for – ships camouflaged in this way.

❖ ❖ ❖

'Battleship grey' is something of a misnomer because it implies an absolute consistency of colour. That was probably not the case across navies or even with the fleet of a single navy. In 2015, Jef

Maytom, an expert involved in the restoration of HMS *Caroline*, which was deployed in the Battle of Jutland in 1916, found thirty-eight layers of paint on the British cruiser's bridge. The 'greys' varied from dark grey to beige-cream.

Design critic Stephen Bayley, writing in the *Guardian* in 2009, made reference to a 'FTSE-100 chairman who went to meetings with a smear of brown dye running from his comb-over down his sweaty forehead, quite unconscious of his comedic absurdity'. (Donald Trump's lawyer, Rudy Giuliani, famously had a similar meltdown when contesting the 2020 elecion results.) As Bayley saw it, any man who dyes his grey hair may

An advert for Wyeth's Sage Tea and Sulphur Grey hair dye, 1925.

as well publicly announce: 'I have disabling psycho-sexual problems about ageing.'

Bayley's view is echoed by Nicolas Sarkozy, who, after he lost the French presidency to François Hollande in 2012, asked a visitor: 'Have you seen him, that ridiculous little fat man who dyes his hair? Do you know anyone like that – a man who dyes his hair?' Edo, a stylist at the Parisian hair salon Les Dada East, believed Hollande's hair colour was deliberate: 'It looks like his politics – austere! He coloured his hair as soon as he came to power. Political men always do that to show they're powerful.'

Though it later turned out that the Socialist president's hairdresser was being paid around €10,000 a month for his services, Hollande himself stated firmly but humorously that his hair colour was natural: 'With all the rain that's fallen on me, my dye would have run.' In the case of Silvio Berlusconi, however, there's no argument. The Italian politician's combination of dyed-black hair, dazzling white teeth and orange tan might be ridiculous, but he's never pretended that it's not artificial. He once sent a bottle of his favourite hair dye to cheer up Egypt's troubled president Hosni Mubarak.

Although grey hair (and/or beard) has long been a signifier of mature wisdom, the fashion for black hair – whether natural or not – has an equally long pedigree. As Rebecca Guenard wrote in the *Atlantic* magazine in 2018: 'Ancient Egyptians dyed their hair but rarely did so while it was on their heads. They shaved it off, then curled and braided it to fashion wigs to protect their bald heads from the sun. Black was the most popular colour until around the twelfth century BCE, when plant material was used to colour the wigs red, blue, or green, and gold powder was used to create yellow.' The Romans produced black dye by fermenting leeches for two months in a lead vessel, a cumbersome process that would seem to indicate some commercial demand.

In the twentieth century, grey's image was altered by the cult of youth created by the entertainment-cosmetics industrial complex. In the 1960s, American inventor Ivan Combe began distributing a haircare product called Grecian Formula under licence. Originally concocted by a Greek barber to cure dandruff, the formula was discovered to gradually turn grey hair brown. In the 1980s ,Grecian Formula was advertised on TV with a slogan cunningly crafted to exploit male insecurity: 'I think the gray's going. Slowly. Gradually. And no one is noticing.' Regulators in Canada and Europe did notice that the formula contained lead acetate, forcing Combe to devise a new formula.

When Manchester United were losing 3–0 away to Southampton in April 1996, Sir Alex Ferguson ordered his players to change out of their grey shirts and into their blue and white alternative strip. The switch didn't exactly inspire a turnaround – United still lost 3–1 – but it may have averted a rout. United's fearsome manager blamed Umbro's kit, insisting: 'The players don't like the grey strip. They said it was difficult to see their team-mates at distance.' United never wore the grey away kit again. (The colour curse struck United again in 2021, when lockdown meant no crowds at football games and Old Trafford was covered in red tarpaulins. After a poor run of home form, the manager, Ole Gunner Soksjaer, said the players found it hard to distinguish each other and had the tarpaulin changed to black.)

Football mythology is full of 'cursed kits', yet research by the University of Durham's Russell Hill and Robert Barton into

Olympic sports in which competitors had been randomly assigned red or blue does suggests that some colours are more
successful than others. Their analysis of four combat events –
boxing, taekwondo, Graeco-Roman wrestling and freestyle
wrestling – at the 2004 Athens Olympics found that, in contests
that were considered evenly matched, contestants in red won
more than 60 per cent of the time.

Hill and Barton could not conclusively explain red's success.
The colour may intimidate an opponent or, as they suggested,
boost the wearer's testosterone levels: 'Maybe you get a surge
when you pull on that red shirt.' If grey – particularly the light
shade worn by United – does affect the outcome, does it have

A lesson learned. When Manchester United turned grey, even Giggs and Beckham
couldn't save them against Matt le Tissier in dominant red.

less to do with spotting a teammate than how the athlete feels about the colour they are wearing? Or, even more importantly, how their opponents feel about it. Quite possibly. A 2008 study, led by Professor Richard Thelwell at Portsmouth University, found that goalkeepers believed that players wearing red would take better penalties than those wearing white and that they had, therefore, less chance of saving them.

Polls suggest that 43 per cent of Americans who believe they have been abducted blame say 'grey' aliens did it. In 2004, the editors of the *Oxford English Dictionary* considered adding these words to its entry on grey: 'A member of any of various supposed species of grey-skinned, humanoid extra-terrestrial beings'.

Many of us grew up envisaging aliens as 'little green men'. The first known appearance of this trope occurs in a short story called 'Green Boy from Hurrah', published in the *Atlanta Constitution* newspaper in 1899. (Hurrah was the name of the green boy's home planet.) By the 1910s, green Martians were featuring in Edgar Rice Burroughs' science fiction and, in pulp sci-fi magazines from the 1920s onwards, the likes of Flash Gordon strove to save the world from fiendish green extraterrestrials. By the middle of the century, however, grey aliens had become part of the cultural zeitgeist. This partly reflected the colour's ambiguous, inscrutable quality, but also drew on the legends that aliens of this colour had been sighted at Roswell, New Mexico, in 1947.

Fourteen years after Roswell, 'greys' were accused of masterminding the first widely publicised alien abduction. On the night of 19 September 1961, Barney and Betty Hill were driving to their home in New Hampshire when a silent, pancake-shaped craft flew right in front of their car. Inside the ship were

Barney and Betty Hill describing their abduction by grey aliens in 1961.

humanoid beings, Betty recalled in a dream, with 'bluish lips …
and skin of a greyish colour'. These aliens then abducted them.
She asked the alien commander where they were from and later,
under hypnosis, she drew the 'star map' she had been shown.
Amateur astronomer Marjorie Fish calculated that the aliens had
come from Zeta Reticuli, a star system some thirty-eight light
years from Earth. True believers argue that we should call them
'reticulans', because that is how they describe themselves, and to
refer to them as 'greys' is racist.

THE WHITENESS OF BEING

'White is not a dead silence, it is pregnant with possibilities.'
Wassily Kandinsky

I s white a colour? If we describe colour purely on the basis of physics, white is not a colour at all. The human eye can see wavelengths of light between 380 and 750 nanometres. The visible spectrum starts with violet (around 380–450 nm) and moves on to blue, green, yellow, orange and finally red (590–750 nm). As white, black, pink and brown do not appear in that spectrum, you could argue that they are not true colours at all. How is it, then, that we can actually see them? Because our eyes mix different wavelengths of light: we see white when all wavelengths of light are reflected off an object and we see black when very few wavelengths are reflected. In other words, white is all colours in one.

In the beginning – not quite the very beginning, but around 100–125 million years ago – there was a flower and it was, as Lewis Dartnell observes in his book *Origins* (2019), 'probably white'. These flowers are known as angiosperms ('encased seeds'), to distinguish them from gymnosperms ('naked seeds'), which developed into all the evergreen conifers today. Darnell writes, 'the angiosperm plants and their pollinators developed together – one of the greatest stories of co-evolution in the history of life on Earth – the world exploded in a profusion of floral colours and heady scents.'

Different pollinators prefer different colours. In a 2009 study at the University of Santa Barbara, researchers found that hummingbirds preferred to pollinate red columbine flowers, whereas hawk moths were more likely to pollinate white and yellow ones. Other studies reinforce the view that, although certain insects gravitate to certain flower colours – experiments show, for example, that coleopterans like white and cream – they can also be influenced by availability, seasonal variations in climate, the number of petals and whether the reward is pollen, nectar or a bit of both.

TB1 – aka 'toilet bowl white' – is the shade that many Americans desperately want their teeth to be. As dentist Ronald Perry told *Nautilus* magazine in 2015: 'What was once considered natural white is now yellow to people. In our society, it's perceived that whiter and brighter is better.'

The Western world's obsession with being 'whiter than white' is reflected in the slogans used to sell washing powder, fridges, soap and dishwashers, and in the architectural style of Charles-Édouard Jeanneret-Gris (1887–1965), better known as Le

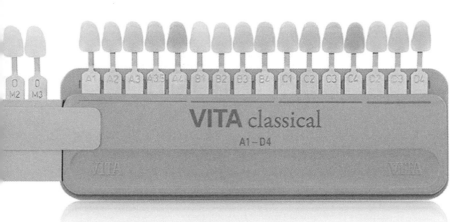

Dentists use shade guides like this to match colours for implants and dentures, but the range doesn't contain shades white enough for modern celebrity teeth.

Corbusier, whose advocacy of white bordered on fanatical. This is how he extolled the morally uplifting power of the enamel paint Ripolin: 'Imagine the results of the Law of Ripolin. Every citizen is required to replace his hangings, his damasks, his wallpapers, his stencils, with a plain coat of white Ripolin. His home is made clean. There are no more dirty, dark corners ... Then comes inner cleanness, for the course adopted leads to refusal to allow anything which is not correct, authorised, intended, desired, thought-out: no action before thought. When you are surrounded with shadows and dark corners you are at home only so far as the lazy edges of the darkness your eyes cannot penetrate. You are not master in your own house. Once you have put Ripolin on the walls you will be master of your own house.'

Le Corbusier's case for the mortal virtues of cleanliness had been made, in the eighteenth century, by John Wesley, founder of Methodism, who proclaimed, in a sermon in 1778, that 'Slovenliness is no part of religion. Cleanliness is indeed next to godliness.' One hundred years later, Procter & Gamble tapped into that belief by launching inexpensive white soap bars. The colour, officially ivory, distinguished their new product from the brown, green and grey soaps already available and, by implying purity and cleanliness, became hugely successful.

Physical cleanliness was a matter of profound concern to officials at the All England Lawn Tennis Club in the 1880s who decided that the best way to minimise unsightly and uncouth sweat stains was to make players wear white. Some Wimbledon champions – notably Pat Cash and Roger Federer – have criticised the rules but when they were updated in 2014 the authorities stressed that white meant pure white, not off-white or cream.

In the West, because we are increasingly likely to see base white as a yellowish white, companies are now often giving their whites a bluish tint – 'bluish white is considered, psychologically speaking, cleaner,' American colour scientist Renzo Shamey told *Nautilus*. Some whitening toothpastes now contain blue covarine, a pigment that sticks to the surface of the teeth to create the illusion of enhanced whiteness. Dentist Ronald Perry is convinced that we will never reach peak whiteness: 'The social and cultural messages are so strong I can't see an end to it.'

How white are Jesus's robes on the Mount of Transfiguration when his divine glory is revealed to three awestruck disciples (James, John and Peter)? Matthew describes them as 'white as light', Luke calls them 'dazzling' and Mark says they became 'radiant, intensely white, as no one on earth could bleach them'.

White is here a symbol of purity and divine perfection. It is simultaneously no colour and all colours – like God, it encompasses everything.

White habits are worn by various Christian orders, including the Cistercians (known as the White Monks), Carthusians, the Dominicans, the Sisters of the Annunciation of the Blessed Virgin Mary and the Trinitarians. Popes have worn white since

St. Bernard of Clairvaux (1090–1153) by the sixteenth-century Spanish painter, Juan Correa de Vivar.

1276, when Innocent V was elected and set a precedent by continuing to wear his white Dominican habit.

The dress that is popularly supposed to have started the Western tradition for brides to wear white – the one worn by Queen Victoria when she married Prince Albert on 10 February 1840 – was actually ivory, rather than pure white, but it nonetheless set a trend. The shift to white, as Summer Brennan observed in a 2017 article for *JSTOR Daily*, also owed a lot to the fact that the 'dresses looked good and stood out in the sometimes muddy-looking new black-and-white or sepia-toned photographic portraits'. By the end of the 1840s, Brennan notes, women's magazines were rewriting history, declaring that white, as an emblem of purity and innocence, had always been the most appropriate colour for a bride.

White's other-worldliness has made it a natural choice for those who wish to proclaim their disdain for materialism. In the 1960s, John Lennon used white to symbolise purity and peace, but he was astute enough about his image to know that he looked good in it. The white denim jacket he wore during a Beatles tour of America was a harbinger of more eye-catching fashion statements to come. In 1969, he was photographed, alongside Yoko Ono, in white high-waisted trousers, white turtleneck jumper, white double-breasted blazer and white pumps. Creating a cult of whiteness, John and Yoko climbed into a white bag for thirty minutes in a gathering of underground artists at the Royal Albert Hall in 1968 and staged various headline-hitting Bed-Ins for Peace in hotel rooms while clad

John and Yoko promoting all-white Bagism together, 1969.

in white. Lennon gave Yoko a white Steinway piano for her birthday, on which he wrote the utopian anthem 'Imagine' in 1971. He later described the song as a sugar-coated version of Marx and Engels' *Communist Manifesto*.

When did white become the standard colour for Modernist architecture? Some say the trend began with the twenty-one buildings of the Weissenhof housing estate in Stuttgart, which was created in 1927 for the Deutscher Werkbund exhibition. The estate was the work of seventeen pioneering architects,

Weissenhof – the fount of all white Modernism ... with a pink tinge.

under the supervision of Mies van der Rohe, who awarded Le Corbusier the commission for the two most prominent sites. *Weissen* means 'to whiten' in German, and nearly every exterior was white. Eleven of the buildings are still standing, and their clean lines and blazing whiteness still pack a punch. Mies van der Rohe's House 1–4, however, has a distinctly pinkish tinge, just as it did originally.

What if paint could cool a building down so much it wouldn't need air conditioning? Engineers at Indiana's Purdue University are convinced that this could happen. They have developed a white outdoor paint that reflects 95 per cent of sunlight, which is considerably higher than any heat-rejecting paints currently on the market. The new paint achieves an effect called 'passive radiative cooling', which means that the temperature of a Purdue-painted surface can be lower than that of the surrounding air – in field tests, the difference was nearly 2 °C in midday sunlight.

The cooling does not consume any energy and the paint reflects sunlight in a way that does not heat up the atmosphere.

'We are redefining and we are restating our socialism in terms of the scientific revolution … The Britain that is going to be forged in the white heat of this revolution will be no place for restrictive practices or for outdated methods on either side industry.'

Harold Wilson, addressing the Labour Party Conference in Scarborough, October 1963

Harold Wilson's most famous political speech cast Labour as the champions of a bold, new technologically empowered Britain. As he said later, in a dig at his Conservative opponents, 'We live in a jet set age, but we are being governed by an Edwardian establishment mentality.' One trade union leader gloated: 'Harold's captured science for the Labour Party.'

Selling himself as a political Modernist, Wilson co-opted white, the dominant colour of Modernist architecture, which was also associated, in the early 1960s, with jet planes, space exploration (the Atlas rocket which launched John Glenn's *Friendship 7* capsule into orbit in 1962 was gleaming white), the PVC raincoat and hat designed by Mary Quant, unveiled the same year as Wilson's speech.

'White heat' is one of the most famous phrases in British political discourse since the Second World War and gave Dominic Sandbrook the title for his bestselling history of the Swinging Sixties. The phrase spun off 'white hot', the colour steel turns when heated to a certain temperature (steel was an industry of immense strategic, social and political import for Labour). One crucial element of Wilson's progressive electoral coalition was the proverbial 'ICI man-in-the-white coat' and his rhetoric suggested that these boffins would help to make it 'physically possible … to conquer poverty and disease'. Sadly, it didn't quite turn out like that, although Wilson's defenders argue that his stalled scientific revolution did stimulate Britain's knowledge economy, partly through the launch of the Open University, sweeping education reform and the creation of a new Ministry of Technology (which Edward Heath effectively abolished in 1970).

The Velvet Underground had a very different kind of 'White Light/White Heat' in mind when they recorded the title track for their 1968 studio album. Lyricist and vocalist Lou Reed said in a 1971 interview that the song was about amphetamines, but Richie Unterberger, in his 2009 book on the Velvets, says 'an equally likely, and perhaps more interesting, inspiration is the American theosophist Alice Bailey's occult book *A Treatise on White Magic*. It advises control of the astral body by a "direct method of relaxation, concentration, stillness and flushing the entire personality with pure white light.' The treatise, which Reed called an 'incredible book' in a radio interview in 1969, includes instructions on how to call down a stream of pure white light.

Until the late nineteenth century, doctors typically wore black for consulations with their patients, to impart the appropriate level of formality to the proceedings. Things changed with Joseph Lister's pioneering research into antisepsis, which proved the paramount importance of cleanliness. The uncontaminated white coat became the signifier of exemplary hygiene, and gave the doctor something of the aura of the laboratory scientists whose rigorous work was displacing much of the quackery that had previously passed for medical knowledge. In hospitals especially, the white coat became the uniform of the profession, as shown in Thomas Eakins' painting *The Agnew Clinic* (1889), where a surgeon in a white smock presides over his white-clad assistants.

The white coat still has an iconic status in American healthcare – the ritual of a 'white coat ceremony' to welcome students into the medical family has become increasingly popular since it was initiated in 1993. However, the traditional long-sleeved white coat disappeared from UK hospital wards in the first decade of this century, when it became apparent that the cuffs were a major cause of cross-infection. Any doctor examining or treating a

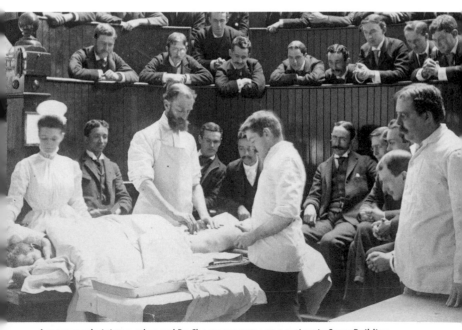

A surgeon administers ether and Dr. Cheever operates on a patient in Sears Building amphitheater, c.1880.

patient in the UK has to be bare-skinned below the elbow. And very few GPs wear white coats any more, partly to prevent 'white coat syndrome' – the sudden elevation in a patient's blood pressure when they see a medic in a white coat.

The Beatles' ninth studio album, officially entitled *The Beatles*, but universally known as the *White Album*, was released on 22 November 1968 on their Apple label. The group's previous album, the psychedelic masterpiece *Sgt. Pepper's Lonely Hearts Club Band*, had left John Lennon anxious to 'just get back to basic music', an intention that was reflected in the songs, the sound and the minimalist title of the follow-up. The idea of calling it *A Doll's House* was dropped after artist Richard Hamilton suggested a white album cover. Hamilton's all-white, double-gatefold cover was a visual palate cleanser after the technicolour glory of Peter Blake's *Sgt. Pepper* artwork. In the original pressing, there were only two marks on the cover: 'The Beatles' embossed in white on the right-hand side of the album and, for the first two million copies, a serial number in the bottom right-hand corner. A fan later paid $79,000 for Ringo Starr's personal copy, number 0000001.

White became the distinctive colour of Apple Computers in 2001, when Jonathan Ive, then chief designer for the company, recommended the 'shocking neutrality' of white for the iPod and subsequent products. Most of Ive's major projects as a British design student had been in white plastic (in 1988, he won a prize for a white landline phone called the Orator),

and he believed that the clarity of white was perfect for Apple's innovative products. Yet, when Ive first presented an Arctic White keyboard to his boss, Steve Jobs said: 'I hate it.' He may have felt the colour was impractical – computers can get grubby pretty quickly – but Ive and his team persevered, developing a spectrum of greys so pale they looked white and giving them cool names such as Moon Grey III and Seashell Grey. By the time Jobs died on 5 October 2011, white had become the predominant colour at the company's new headquarters, Apple Park, in Cupertino, California.

Apple played a significant part in making white the coolest colour of the early twenty-first century. As a Ford design manager told *Consumer Reports* in 2014: 'You might think that white is just a boring colour that is never going to change but Apple helped that trend move on.' White paint makes cars literally cooler by reflecting light and heat. With new technology, the detailing on white cars improved. Economic factors were at work, too – the credit crisis of 2008 and the ensuing financial downturn prompted car buyers to prefer conservative colours, with greyscale tones (white, black, grey, silver) especially popular. And white was often the cheapest colour to manufacture, and thus the least expensive to buy. In 2019, a little under 40 per cent of the cars sold worldwide were white.

The 'white van man' – most likely to be a self-employed builder, plumber or electrician – came to prominence at the turn of the century, being simultaneously demonised as the bane of all other motorists and celebrated as entrepreneurial working-class heroes by such tabloids as the *Sun*. The phrase was first used on 18 May 1997 by Jonathan Leake, transport editor of the *Sunday Times*, who described them as the 'scourge of the road'. Yet politicians who sneer at them jeopardise their careers. In 2014, Emily Thornberry had to resign from Labour's shadow cabinet after tweeting a photo of a white van outside a house with English flags draped over it, with the caption: 'Image from #Rochester.' This picture, she implied, epitomised bone-headed English nationalism.

The shade is officially called Navajo white, but most of us would call it pale orange. The colour, which graces the background of the Navajo Nation flag, was popular with interior designers in the 1970s, who used it in low-budget housing because it doesn't show up the dirt. Confusingly, the Navajo, like other Native Americans, regard plain old conventional white as one of the four sacred colours that correspond to cardinal directions and a time of day: white signifies east and the dawn; light blue, south and day; yellow, west and dusk; and black, north and night. In Navajo myth, creation happened in four stages or worlds. The fourth world, in which the First Man and First Woman became human, is known as *Nihalgai*, the white or glittering world.

Procol Harum's signature song is one of the few numbers in the rock pantheon with a title that sounds like something from a paint catalogue: 'A Whiter Shade of Pale'. Keith Reid, who wrote the much-analysed lyrics, says he heard the title phrase at a party – 'Some guy looked at a chick and said "You've gone a whiter shade of pale"' – and wrote a song about it. Fascinated by the European art movies *Last Year in Marienbad* (1961) and *Pierrot le Fou* (1965), he decided to tell his girl-leaves-boy story like scenes in a movie. The dreamlike lyrics prompted rumours (which Reid vehemently denies) that it was about drugs. As Mike Butler noted in the anthology *Lives of the Great Songs* (1994), it is more likely to be about booze and sex – 'Her face at first just ghostly, turned a whiter shade of pale' is one of several allusions to nausea in the lyric. Procol Harum devotees might have felt a bit nauseous themselves in 1982, when the song, rearranged for pan-pipes, featured in a sickly TV advert for Dulux.

The White Lady is the name of a cocktail of crème de menthe, Cointreau or Triple Sec, fresh lemon juice and (sometimes) an egg white. It was probably invented by Harry MacElhone at Ciro's Club in London in 1919 and tweaked by him at Harry's Bar in New York, ten years later. There is a counter-claim that Harry Craddock created it at the American Bar at the Savoy.

The appearance of a ghostly white lady is seldom good news. Perhaps because so many of them are believed to be the spirits of women who were murdered, committed suicide or had lost a husband or child, their appearance in a household or – more spookily still – in a photograph has been widely believed to signify an imminent death in the family.

In Malta, the White Lady of Mdina lures heart-broken teenage boys and elderly men to join her ghostly band of followers. In eastern and northern Netherlands, white ladies are reputed to switch newborn babies, whereas in Quezon City, in the Philippines, they like to terrify taxi drivers. The white ladies which haunt the towns of Beeford (East Yorkshire), Chicago (Illinois), Mukilteo (Washington) and Altoona (Pennsylvania) are more restrained, usually manifesting themselves as vanishing hitchhikers.

Generally regarded as one of the earliest examples of the mystery novel, Wilkie Collins' *The Woman in White*, first published in serial form in 1859–60, was such a huge commercial success it inspired a trend for white bonnets and cloaks, and prompted perfumiers to concoct Woman in White scents. Composers even turned out Woman in White waltzes and quadrilles.

At the start of the book, Walter Hartright, in London, comes to the aid of a mysterious and distressed woman who is dressed in white. A very similar episode, occurring a few years earlier, is related in *The Life and Letters of Sir John Everett Millais*, by the

John Everett Millais' *The Somnanbulist*.

artist's son, John Guille Millais. Late at night, Wilkie Collins and his brother Charles were accompanying the painter back to his house when 'they were suddenly arrested by a piercing scream coming from the garden of a villa close at hand'. A few moments later, 'the iron gate leading to the garden was dashed open, and from it came the figure of a young and very beautiful woman dressed in flowing white robes that shone in the moonlight. She seemed to float rather than to run in their direction, and, on coming up to the three young men, she paused for a moment in an attitude of supplication and terror. Then, seeming to recollect herself, she suddenly moved on and vanished in the shadows cast upon the road.'

It's possible that the incident was exaggerated or even fabricated, but whatever the truth of the matter it's interesting that Millais, as well as his friend Collins, went on to create an enigmatic woman in white: Millais' *The Somnambulist*, which he completed in 1871, might be the sibling of the muslin-clad young woman who enchants Walter Hartright.

The most famous nineteenth-century image of a woman in white is a picture painted by James McNeill Whistler in 1862. Showing his mistress Joanna Hiffernan standing in front of a white muslin curtain, it was originally entitled *The White Girl*, then *Symphony in White, No. 1*, and was first displayed in the Berners Street Gallery, where it was advertised as *The Woman in White*. The notoriously combative artist took exception to this, writing in a letter to the editor of the *Athenaeum*: 'The proprietors of the Berners Street gallery have, without my sanction, called my picture "The Woman in White". I had no intention whatsoever of illustrating Mr. Wilkie Collins' novel. It so happens, indeed,

Whistler's *Woman in White: Symphony in White No. 1.*

that I have never read it. My painting simply represents a girl dressed in white standing in front of a white curtain.'

Was Whistler protesting too much? It's likely that he was happy to stir up some controversy to raise his profile. Indeed, Frederick Buckstone, the secretary of the Berners Gallery, felt obliged to inform the *Athenaeum* that 'Mr. Whistler was well aware of his picture being advertised as "The Woman in White" and was pleased with the title.'

Whistler was only twenty-seven when he created *Symphony in White, No. 1*, and he hoped it would make his name. Initial reaction, though, was not entirely favourable. Some critics thought the young woman's blank expression was unladylike, while another reviewer complained that 'her anomalous white garment … hangs upon her person in absolute defiance of all ordinary canons of good taste'. Having been spurned by the Royal Academy, in 1863 he submitted the painting to the Salon in Paris, the biennial showcase of the Académie des Beaux-Arts, which also rejected it. It was accepted, however, by the Salon des Refusés, an exhibition set up by Gustave Courbet for work turned down by the Salon. There it was rather overshadowed by Manet's scandalous *Le Déjeuner sur l'herbe*, but was admired by some heavyweight connoisseurs, among them Courbet, Manet and Baudelaire. Whistler had arrived. He went on to paint a *Symphony in White, No. 2* and a *Symphony in White, No. 3*, both of them featuring Joanna Hiffernan.

Symphony in White, No. 1 was more than two metres tall and featured copious amounts of lead white. Whistler later maintained that his exposure to the pigment in the creation of his masterpiece had made him ill – so-called 'painter's colic', an illness that affected the nervous, digestive and respiratory systems, was known to be a form of lead poisoning. Whistler did retire to Biarritz to recuperate from an illness after finishing the painting, but it seems unlikely that a single painting could have severely poisoned him. Which is not to say that his health wasn't compromised by lead white – he used copious amounts of it throughout his career.

The pigment's density, opacity and warmth had beguiled earlier artists such as Vermeer and Rembrandt, who perhaps regarded lead poisoning as an occupational hazard. Fortunately, this beautiful but toxic tint has been replaced by titanium white. It isn't poisonous, but as Victoria Finlay noted, 'it doesn't quite have the wicked twinkle of lead'.

'At first, you're like "Why are they stealing the colour white?"' mused Dean Chappell of the FBI to *Business Week* reporters in 2016. There were, he soon realised, 2.6 billion reasons why someone would want to steal DuPont's secret formula for titanium dioxide (TiO_2), which was the basis for a division generating $2.6 billion in revenue.

A natural oxide, usually extracted from ilmenite ore, TiO_2 was first used as a pigment in the nineteenth century. In the 1940s, DuPont refined its own version, widely regarded as superior to anything else on the market. The company does not talk publicly about how it makes its TiO_2, but *Business Week* reports that this extremely complex process uses chlorine, carbon, purifying chemicals, low temperatures, heat and oxygen to produce particles that are so fine they have the consistency of talcum powder.

TiO_2 is used to whiten everything from the hulls of superyachts, sheets of paper and the chalk lines on tennis courts, it was certainly worth stealing – which is what Walter Liew, a naturalised American citizen, business owner and technology consultant, did from 1997 to 2011, with help from retired DuPont engineer Robert Maegerle. The trade secrets – including a blueprint for DuPont's new TiO_2 factory – were then sold to companies controlled by the Chinese government for $20 million. In 2014, Liew was sentenced to fifteen years in prison.

Although the term 'white noise' is often used to describe any discordant background noise, scientifically it is defined as a sound containing an equal amount of every audible frequency. Just as white light contains all the colours we can see, white noise is the sum of all the sounds we can hear. In its purest form, as Meghan Neal wrote in the *Atlantic* in 2016, 'white noise sounds like that hissy "shhh" that happens when the TV or radio is tuned to an unused frequency'.

The uniformity of white noise can mask the sudden changes or inconsistencies in sound that wake many of us up at night. White noise generators have become a common sleeping aid, although studies have shown that brown noise – in which the frequencies are changed so the lower notes sound more powerful – may be just as useful in helping us sleep deeply.

Belgian sound engineer Stéphane Pigeon says that white noise can help our ears to become more discerning. 'Before it was just the sound of running water, while now I realise that every river has its own sound. I feel I can hear things I didn't hear before,' he says. 'I'm listening to everyday sound but with a different ear.'

❖ ❖ ❖

Whitewashing a house with white paint or chalked lime was cheap, but – because it flaked – the wash had to be reapplied every year. The social nuances of this choice inspired the Kentucky adage 'Too proud to whitewash, too poor to paint.' Although American schoolchildren have long been told that the White House was painted white to hide the burn marks left after the British set fire to it in 1814, the American government had

The White House – as it appeared on old dollar bills.

been whitewashing the grey sandstone walls since 1798 to waterproof them and stop them cracking in winter. In 1818, white lead paint was used for the first time. By that time, the residence was already known as the White House, but the name was not made official until October 1901, when Theodore Roosevelt moved in.

In Britain and America, whitewash suggests a cover-up or intent to deceive. The term 'whitewash' was applied to John Adams, the second American president, although in this instance, the *Philadelphia Aurora* newspaper was calling on his supporters to stop the Democrats blackening his reputation. By 1973, with Watergate wrecking his presidency, Richard Nixon used the term in the opposite sense, declaring: 'There can be no whitewash at the White House.'

The gleaming white stone of ancient Greek buildings and sculpture is misleading – at the time of their creation, they would have been painted in various shades of blue, red, yellow and many other colours. The Parthenon, for example, used to house Phidias's huge statue of the goddess Athena, which the second-century historian Pausanias described as 'chryselephantine, covered in gold and ivory'. Victorian artist Lawrence Alma-Tadema was probably not too far off the mark in his painting *Phidias Showing the Frieze of the Parthenon to his Friends* (1868), which depicts the temple in dazzling red, white and gold. Such colour schemes found no favour with Auguste Rodin, who is purported to have banged his chest and declared: 'I feel it there that they were never coloured.'

Sarah Bond, an Assistant Professor of Classics at the University of Iowa, has argued that the mistaken belief that the sculpture of the classical world was overwhelmingly white has encouraged the assumption that white skin was the classical ideal. It does not help, she said in *Forbes* magazine in 2017, that museums and art history textbooks tend to present 'a predominantly neon white display of skin tone when it comes to classical statues and sarcophagi ... [this] assemblage serves to create a false ideal of homogeneity – everyone was very white! – across the Mediterranean region. The Romans, in fact, did not define people as "white" so where, then, did this idea of race come from? The equation of white marble with beauty is not an inherent truth of the universe.' She was trolled – and threatened – online for her views.

Not so Classical White – Alma-Tadema's probably historically accurate imagination of *Phidias Showing the Frieze of the Parthenon to his Friends*.

From the nineteenth century onwards, as more people (mainly men) began working in offices, white shirts entered the dress code. By the 1920s, this code was so prevalent that the American novelist Upton Sinclair called these people 'white collar workers', distinguishing them from manual labourers, who tended to wear darker colours – often blue boiler suits with blue collars – which hid the dirt better. In this context, the pursuit of 'whiter than white' shirts, tablecloths and sheets, encouraged by the invention of Persil, a soap powder with a bleaching agent, in 1903, can be partly interpreted as a desire to distinguish oneself from the lower echelons of society. You might not be able to find a white-collar job, but you could have a shirt with a 'whiter than white' collar and dazzle your relatives with the purity of your tablecloth. Such coded communication is still with us – as Kassia St Clair notes in *The Secret Lives of Colour*: 'Someone wearing a snow-pale winter coat telegraphs a subtle visual message: "I do not need to use public transport."'

French composer Erik Satie claimed, in his autobiography, to dine only on 'food that is white: eggs, sugar, shredded bones, the fat of dead animals'.

Nobody knows what killed Elizabeth I. On her orders, no post-mortem was conducted – possibly to safeguard her reputation as a virgin. Pneumonia, streptococcus and cancer have all been suggested, but she may have been poisoned by the white lead in her make-up.

Elizabeth I in her coronation robes, patterned with Tudor roses and trimmed with ermine; seventeenth-century painting by an unknown artist.

The queen's determination to preserve the mask of youth was partly vanity – at the age of twenty-nine she caught smallpox, which marked her skin – but as Tracy Borman, joint chief curator of Historical Royal Palaces, writes, 'Any outward sign of infirmity on a sovereign's part undermined the immortal, God-like status that was essential to retaining their power.' To create her famously pale complexion, it is said that Elizabeth's ladies applied a cosmetic called Venetian ceruse, which contained white lead and vinegar, and then glazed her face with egg white. Victims of lead poisoning often suffer from rotting teeth and hair loss – and we know that Elizabeth was bald and had black teeth by the time she died, at the age of 69, on 24 March 1603.

Borman says there is evidence that she began losing her hair at the age of thirty.

Some historians have argued that descriptions of the decrepit monarch are misogynist fictions. This is the view of Kate Maltby, who also points out that Venetian ceruse was never found in the queen's inventories. Helen Hackett has argued that the image of Elizabeth I as 'a decaying hag consumed by vanity' is a Victorian invention, stressing her sterility and unfemininity by way of contrast with Queen Victoria, 'a very fertile, fecund figure as a wife and mother'. Be that as it may, Tracy Borman is certain that the queen 'underwent a lengthy and painstaking beauty regime every single day of her forty-four-year reign'. Even if this regime did not involve Venetian ceruse (which was a hugely popular skin whitener during her reign), it almost certainly contained cosmetics that included enough lead white to cause damage.

The fashion for pale faces outlived Elizabeth. White skin was a symbol of social rank, signifying that someone either didn't work at all or didn't labour in the open air. Aristocrats and social climbers of both genders painted their faces with lead white. Being described as 'dead white' was a compliment. In *A History of Make-Up* (1970), Maggie Angeloglou quotes an eighteenth-century advert for a chemical wash that would improve the skin 'by taking off all deformities … as Ringworms, Morphew, Sunburn, Scurf, Pimples, Pits or Redness of the Smallpox, keeping it of lasting and extreme Whiteness'.

At times of crisis – real, perceived or invented – Kim Jong-un is sometimes depicted riding a white stallion through snow on

Mount Paektu, the mythical birthplace of the Korean nation. The photo opportunity is designed to evoke Chollima, a winged steed in classical Chinese mythology (although the horse, which legend has it, can be ridden by no mere mortal, was not necessarily white). 'Chollima' was the official name given to North Korea's economic recovery plan in the 1950s, is the nickname of the national football team and has featured on the country's currency and stamps.

In a land where Communism, nepotism and religious mythology intertwine, Kim's propaganda exercise underlines his descent

The Great Leader, Kim Jong-un, mounted on Chollima.

from – and proximity to – his grandfather, Kim Il-sung, the founder of North Korea. In his memoirs, Kim Il-sung describes a pure white horse as his trusted ally during the Korean revolution. In 2019, North Korea spent $75,000 importing a dozen white horses and ponies from Russia. Kim Jong-un sometimes gives white horses to members of his inner circle.

A miraculous white horse is to be found in many ancient cultures. Xerxes I, the Persian emperor who reigned from 485 to 465 BCE, is said to have kept a stable of sacred white horses. In other cultures, they are associated with fertility, warrior heroes, the end of the world (and the advent of the world to come), snow (in Blackfoot mythology, the snow god Aisoyimstan rides a white horse) and rain (in Zoroastrianism). In Bronze Age Britain, the locals expended considerable effort to carve the figure of a white horse into the hills near Uffington, in what is now Oxfordshire.

Pegasus, the immortal winged horse of Greek mythology, is usually depicted as a pure white animal. Blessed with the useful gift of being able to create a spring with a tap of its hoof, and a great support to his owner, the monster-slaying hero Bellerophon, Pegasus set the template for many subsequent white steeds, whose riders have ridden to the rescue of maidens in distress and the oppressed in general – the hero in many B-movie Westerns rides a white horse, as does the Lone Ranger.

In the West, a 'white elephant' signifies as a grandiose project with no value or purpose. The metaphor is often traced back to an (unnamed) king of Thailand who allegedly gave white elephants to people he was displeased with, in the hope that the expense of looking after the beasts would ruin them.

In Thailand itself, as in many Asian countries, the white elephant is regarded as an incarnation of Gautama Buddha. According to legend, the Buddha's birth was revealed to his mother Maha Maya in a dream in which a white elephant with six tusks entered her right side, a sign that she had conceived a child who would rule the world or become a god.

White elephants have been regarded with so much reverence that the kings of Thailand, Cambodia and Burma fought several wars from the sixteenth to the eighteenth centuries in order

A royal white elephant depicted on a traditional Thai painting.

to get their hands on the animals. Bhumibol Adulyadej, king of Thailand from 1946 until his death in 2016, is said to have possessed twenty-one white elephants, more than any of his predecessors.

Albino elephants are rare, and are not really white – they are a soft reddish brown, which looks pink when wet. Pale grey elephants can be classified as white and, though they are not as highly prized as albinos, they are still sacred, and for that reason they cannot be put to work – which is why the gift of a white elephant was more a hindrance than a benefit.

Under a White Sky is the title of *New Yorker* science writer Elizabeth Kolbert's book on the world we may need to create to combat climate change. The white sky is a vision of our possible future if, as many scientists now believe, we have to employ geo-engineering solutions to cool the planet. The most promising but most drastic plan, untested even in a small-scale experiment, because it is too awful to contemplate, is for solar engineers to shoot a million tonnes of sulphur dioxide particles into the stratosphere to reflect sunlight away from the planet. This would have to be done repeatedly until we have somehow removed enough CO_2 from the atmosphere to reverse global heating. One potential side effect is an alteration of the spectrum of light reaching Earth. The sky would turn from blue to white.

ACKNOWLEDGEMENTS

My editor Jonathan Buckley has had more influence on *The Colour Code* than Ezra Pound had on *The Waste Land* and I am proud to acknowledge that he is, as Eliot said of his mentor, 'il miglior fabbro'. My publisher Mark Ellingham has also helped transform this book with his wise counsel, eloquent advocacy and inspired picture editing. This book is dedicated to both gentlemen and to my wife Lesley and son Jack. I would also like to acknowledge my debt to the following people who have, in various ways, improved the book by sharing their insight and talents: Shaul Adar, Sally Augustin, Andy Barton, Dominique Campbell, Bevil Conway, Simon Curtis, Ian Cranna, Steve Cropley. Claudia Daventry, Hunter Davies, Jules Davidoff, Andrew Franklin, Alastair Griffiths (Royal Horticultural Society), Paul Harpin, Kate Hess, Uli Hesse, Henry Iles (for his impeccable design), my late, dear friend Jim Izzard, Natania Jansz, Martin Mazur, Sabine K. McNeill, Mel Nichols, Alison Ratcliffe (for indefatigable, multi-lingual research), Marilyn Read, Stuart Semple, Frank Summers (at the Space Telescope Science Institute), Claudio Tognelli, Nikky Twyman (for assiduous proofreading), Michael Yokhin and Valentina Zanca.

I would also like to acknowledge the brilliance of Philip Ball, Victoria Finlay, John Gage, Kassia St Clair and Michel Pastoureau, whose writings on colour have been an immense, and thoroughly enjoyable, inspiration.

IMAGE CREDITS

Individual credits

Feynman graphic novel – Jim Ottaviani and Lelan Meyrick (0:2 Books) p.19.

Colour visual of Messiaen's mode 33, © Håkon Austbø, from 'Visualizing Visions: The Significance of Messiaen's Colours', Music & Practice (www.musicandpractice.org) p.21.

Morison's Vegetable Universal Pills (Wellcome Collection) p.76.

'Diamond Dan' the Orange Man (Photopress Belfast) p.148.

Kaye Blegvad's graphic novel about depression, *Dog Years* (p.285) is available from kayeblegvad.com.

Alamy

Captain Scarlet p.31; Red Baron p.33; Riley sees red (AF Archive) p.35; Ferrari rally (Shawshots) p.39; Garibaldi and his Reds (Interfoto) p.64; Cochineal dye (Interfoto) p.66; Amanda Gorman (Pool) p.72; Yellow Brick Road (Pictorial Press) p.88; Yves Klein (Vincent West) p.115; Rajneeshis in Oregon (SCPhotos) p.152; Orange Revolution in Kiev (Watchtheworld) p.156; Kabuki actor Kuniya Sawamura p.166; Jim Clark in his Lotus 25-Climax (National Motor Museum) p.201; Arsenic-laced wallpaper (ICP) p.212;Gulabi Gang (Joerg Boethling) p.245; Yohji Yamamoto, armed in black (Nicolas Gouhier/ABACAPRESS) p.281; Megan Fox (dpa) p.301.

Getty

Inter Milan (Jasper Juinen/Getty Images) p.47; Taxi poster p.83; Odysseus enduring the Sirens (Mondadori/Hulton Fine Art Collection) p.104; 'Blue Gold' (Remy Gabalda/AFP) p.107; Hathor (Sepia Times/Universal Images Group) p.112; Boat Race (Archive Photos) p.124; Krishna dancing (Frédéric Soltan) p.126; Bonampak p.128; Ruud Gullit (Paul Popper/PopperFoto) p.146; Perkin's mauve (Science & Society Picture Library) p.171; Han purple (Visual China Group) p.173; Hamish Bowles (Dimitrios Kambouris) p.180; Beirut's green line (Mark DeVille/Gamma-

Rapho) p.195; Elvis Presley (Michael Ochs Archives) p.228; Princess Diana steps out in Stambolian (Tim Graham Archive) p.278; When Manchester United turned grey (Shaun Botterill) p.313; Barney and Betty Hill (Bettman) p.315; John and Yoko (Hulton Deutsch) p.325.

iStock

Moroccan dyes p.2; Jodphur p.110; Chicago River turns green p.205; Kyoto's Maruyama Park p.241; Brown shoes p.257; Ayam Cemani p.289; Titanium dioxide (p.338); White House (p.340).

WikiCommons/Library of Congress/public domain

Moses Harris colour wheel (WikiCommons) p.6; Goethe's colour wheel (WikiCommons) p.9; Chevreul's The Laws of Contrast of Colour (WikiCommons) p.11; Thomas Young's Lectures (WikiCommons) p.12; Mantis shrimp (Silke Baron/ WikiCommons) p.14; 'Blue/gold' dress (Twitter) p.17; Werner's Nomenclature of Colours (WikiCommons) p.23; Steppe bison (Museo de Altamira/D. Rodríguez/WikiCommons) p.28; Victorian Christmas card p.36; Mondrian's Composition II in Red, Blue, and Yellow p.43; Red heifers, Temple Institute (templeinstitute.org), p.41; True Blood sex chart (truebloodonline.com) p.44; Battle of Gennis (www.britishempire.com) p.48; Sans culotte (Library of Congress) p.50; Neue Rheinische Zeitung (Arquivo Marxista na Internet/Wikipedia.de) p.52; Little Red Book (ebay) p.54; red scare comics (Pinterest) p.55; US election 1976 (Pinterest) p.57; Lipstick shades (Pinterest) p.58; Gone with the Wind poster (Pinterest) p.60; Stripe advert (Pinterest) p.61; Botticelli's Venus (WikiCommons) p.70; Van Gogh's Starry Night p.74; Indian Yellow (Winsor & Newton) p.75; Beardsley's Yellow Book (Pinterest) p.78; Private Eye (Private Eye) p.78; Kupka's The Yellow Scale (Pinterest) p.81; Touraine caricature (WikiGallery) p.85; Brazil 1958 p.91; Maillot Jaune (Gallica BNF) p.93; Delacroix's The Execution of Marino Faliero (WikiCommons) p.95; Giotto's The Kiss of Judas (WikiCommons) p.97; Scattering (WikiCommons) p.103; Earthrise (NASA) 109; Gainsborough's Blue Boy (WikiCommons) p.116; YInMn Blue (WikiCommons) p.119; Chagall window at Tudeley (Klaus D. PeterWikiCommons) p.120; Saint Louis

Blues (ABE Books) p.123; Applause rose (WikiCommons) p.132; Rancho Mirage (TopTenRealEstateDeals.com) p.136; Kandinskys Color Study, Squares with Concentric Circles (WikiCommons) p.138; Guru Nanak (WikiCommons) p.140; Monet's Impression, Sunrise (WikiCommons) p.143; Orpiment (crystalage.com) p.144; Orange dwarf (Pnapora/WikiCommons) p.154; Rubens' The Secret of Purple (WikiCommons) p.159; Pompeii fresco (WikiCommons) p.161; Napoleon I by Ingres (WikiCommons) p.163; Prince Shotoku (WikiCommons) p.165; Bee's eye view of colour (openphotographyforums.com) p.175; Monet's Waterliliies (WikiCommons) p. 177; Castel's colour organ, (WikiCommons) p.185; Purple Heart (WikiCommons) p.188; Green Man at Rochester Cathedral (WikiCommons) p. 192; Gawain and the Green Knight (WikiCommons) p.193; Harry Sherman (Pinterest) p.198; Errol Flynn (Pinterest) p.203; Kirchenväter altarpiece (WikiCommons) p. 207; Jack in the Green (Pinterest) p.209; Original greenback (WikiCommons) p.210; Manet's The Balcony (WikiCommons) p.215; Van Eyck's Arnolfini Wedding (WikiCommons) p.219; Gaddafi's Green Book (Pinterest) p.221; Elizabeth I Hampden Portrait (WikiCommons) p.224; Pink pussy hats (Ted Eytan/ WikiCommons) p.226; Tiepolo's Venus and Vulcan (WikiCommons) p.230; Marilyn Monroe (WikiCommons) p.232; Sears Christmas catalogue (Pinterest) p.234; Memorial plaque at Buchenwald (WikiCommons) p.237; MXY-7 Ohka (Pinterest) p.238; Seattle Naval Correction Center(Pinterest) p.242; Stuart Semple's patent pink (Instagram) p.248; Pantone 448C (Pinterest) p.252; Saint Francis (Pinterest) p.253; Pullman Golden Arrow and Orient Express posters (WikiCommons) p.256; Brownie camera advert (WikiCommons) p.259; Rembrandt's Night Watch (WikiCommons) p.261; Vandyke brown (Winsor & Newton) p.262; Jackie Robinson (Smithsonian) p.264; Martin Drolling's Interior of a Kitchen(WikiCommons) p.266; Redemption of Vanity (Diemut Strebe/MIT) p.270; Charles Johnson's A General History of the Pyrates (WikiCommons) p.276; Kali (WikiCommons) p.273; Malevich's Black Square (WikiCommons) p.28l; Black Country, Night, With Foundry by Edwin Butler-Bayliss (Wolverhampton Art Gallery) p.287; Chicago Black Hand note (Chicagology) p.290; Jules Girardet –the arrest of Louise Michel (WikiCommons) p.293;